The ABCs of EMP

A PRACTICAL GUIDE TO BOTH UNDERSTANDING AND SURVIVING AN EMP

JEFFREY YAGO

Dunimis Technology Inc.
Gum Spring, Virginia

The ABCs of EMP

Copyright © 2020 by Jeffrey Yago

All rights reserved. No part of this book may be reproduced in any form or by any means—whether electronic, digital, mechanical, or otherwise—without permission in writing from the publisher, except by a reviewer, who may quote brief passages in a review.

Published by
Dunimis Technology Inc.
P.O. Box 10
Gum Spring, Virginia 23065

Library of Congress Control Number: 2020907111

ISBN: 978-1-7346385-6-1
EBOOK: 978-1-7346385-7-8

Printed in the United States of America

"You don't need a weatherman to know which way the wind blows."
—BOB DYLAN

NOTICE TO READERS

THE MATERIAL IN THIS BOOK is for voluntary acceptance and use by the reader. This book is not meant to define the design or safety standards for any electrical wiring or EMP protection techniques. Any included suggestions and product selections are intended to assist the reader in developing their own self-reliance plans.

All wiring and electrical equipment installed in new or existing construction may be subject to local building codes and must be installed according to the National Electric Code. It is recommended to enlist the assistance of a professional if in doubt.

This material is subject to revision as further experience and product development may deem necessary in this rapidly changing field. Any products referred to by brand name or model number are for informational purposes only and not intended to be an endorsement by this author or publisher.

Contents

Contents 5
Foreword vii
Acknowledgements ix
Introduction 12
Chapter 1: Understanding Electromagnetic Disturbances 18
Chapter 2: What Is an EMP? 29
Chapter 3: EMP Basics 34
Chapter 4: Don't Forget the Sun 45
Chapter 5: EMP Impact on the Power Grid 52
Chapter 6: EMP Inaction by Government and Grid Operators 63
Chapter 7: EMP Impact on Our Military 74
Chapter 8: EMP Effects on Communications 89
Chapter 9: EMP Impact on SCADA 96
Chapter 10: EMP Impact on the Banking Industry 104
Chapter 11: EMP Impact on Vehicles 110
Chapter 12: EMP Effects on Transportation Control Systems 115
Chapter 13: EMP Impact on Oil and Gas Distribution 120
Chapter 14: See Who Is at the Door 124
Chapter 15: So, What Can We Do? 133
Chapter 16: Transportation after an EMP Event 146
Chapter 17: Will Your Generator Survive an EMP? 151
Chapter 18: Lighting after an EMP 155
Chapter 19: Powering Communications after an EMP 162
Chapter 20: Powering Computers after an EMP 172
Chapter 21: Powering Audio/Video Systems after an EMP 179

Chapter 22: Powering Medical Equipment after an EMP...........184
Chapter 23: Powering Security Systems after an EMP 187
Chapter 24: Protecting Battery-Powered Tools from an EMP ...196
Chapter 25: Powering Water Pumps after an EMP.....................200
Chapter 26: Powering Refrigeration after an EMP 206
Chapter 27: Using Vehicles for Battery Charging 214
Chapter 28: Charging Batteries with Solar Power......................218
Chapter 29: Which Rechargeable Batteries? 226
Chapter 30: Connecting Battery-Powered Devices....................236
Chapter 31: Protecting Your Vehicles from EMP.......................243
Chapter 32: Protecting Antennas and Shortwave Radios from EMP 247
Chapter 33: Protecting Home Appliances from EMP 254
Chapter 34: Protecting Solar Systems from EMP 262
Chapter 35: Bugging Out after an EMP 269
Chapter 36: Basic Faraday Devices for EMP Protection 274
Chapter 37: Faraday Rooms for EMP Protection 283
Chapter 38: Closing Comments 290
Appendix 298
Photo Credits..299
Notes 300
Index 307

Foreword

After many years investigating the condition of emergency backup power systems and automated controls for numerous universities, hospitals, government buildings, all NASA facilities, military bases, and headquarters for several "3-letter" agencies, Jeff Yago learned firsthand the lack of backup power readiness in the United States today. Since the early 1980s Jeff was designing off-grid solar power systems when most building owners were just beginning to consider this as just a future possibility. One of his early solar power designs is still powering an off-grid university observatory on a rural mountaintop, which is just one example of his unique design experience.

In 1992 he was inducted into the "Order of the Engineer" by the National Society of Professional Engineers, to recognize those professional engineers who have demonstrated the highest level of professional ethics throughout their engineering career.

Jeff is known for his numerous articles dealing with solar and backup power systems for magazine publications including *Self-Reliance*, *Home Power*, *Backwoods Home*, and *Mother Earth News*. He has been a featured speaker at self-reliance expos throughout the country each year and has discussed his concerns for the lack of preparedness on multiple national radio talk shows including Doug Hagmann "The Hagmann Report;" John B. Wells "Caravan to Midnight;" Jack Spirko's "Survival Podcast;" James Lowe's "Transmedia Worldwide;" and the "Josh Tolley Show."

FOREWORD

I met Jeff after he published his earlier book, *Lights On*, which was ranked highest in sales on Amazon in the self-reliance category. He is known for taking complex subjects and turning them into easy-to-understand reading using his unique writing style. This has served him well when wanting to tackle the very difficult subject of EMP. Most books on the subject appear to have been written for a graduate level physics classroom, while others are over-simplified with no real explanation of how an EMP is actually generated, what it can do, what it can't do, and more importantly how to prepare when it finally happens.

Jeff has spent many years researching all available EMP publications, had endless correspondence with numerous government agencies, and met with experts in the EMP field to finally bring this all together. I hope you will find this an eye-opening, yet interesting read.

> Dr. Peter Vincent Pry
> Executive Director of the EMP Task Force on National and Homeland Security, and Director of the United States Nuclear Strategy Forum

Acknowledgements

During the search for the real truths concerning the EMP debate I was fortunate to meet Dr. Peter Vincent Pry, considered by many to be the dean of the EMP school of thought. Dr. Pry served as a Verification Analyst monitoring Soviet compliance with various arms control treaties for the U.S. Arms Control and Disarmament Agency before moving to the CIA, where he served ten-years (1985-1995) as a senior analyst of Soviet and Russian nuclear weapons and strategy, including serving as CIA's top analyst on EMP. This experience going head-to-head with his counterparts in Russia convinced him the Soviet Union was heavily involved in EMP weapons research, and had operational plans to make nuclear EMP attacks an integral part of their offensive military capabilities.

Dr. Pry organized the first ever series of unclassified hearings on EMP before Congress over the next six years (1995-2001) to educate policymakers and the public about this (at the time) little known existential threat. Dr. Pry helped establish and served as Chief of Staff of the Commission to Assess the Threat to the United States from Electromagnetic Pulse (EMP) Attack (2001-2008 and re-established 2016-2017), which was comprised of the nation's foremost experts on EMP, strategy, intelligence, and critical infrastructure. Over the next seventeen-years they produced the most comprehensive documentation available regarding EMP threat assessment, including recommendations for protecting national critical infrastructure.

Dr. Pry is currently the Executive Director of the EMP Task Force on National and Homeland Security, and Director of the United States Nuclear Strategy Forum, both nonprofits and

ACKNOWLEDGMENTS

unfunded advisory boards to Congress. His book titled *THE POWER AND THE LIGHT: The Congressional EMP Commission's War to Save America 2001-2020* was just published in March 2020 and is a great insider's view of the long, and still ongoing struggle within the US government to protect the American people from the existential threat that is an EMP.

His assistance and guidance during the completion of this book has been invaluable.

During my research, I also met with Bob Goldblum at a 2005 conference in Chicago. Bob was heavily involved with the monitoring of all forms of electromagnetic radiation emissions during the early nuclear bomb tests in Nevada during the 1960s. In the 1970s Bob operated testing facilities in multiple states to test military equipment against all forms of electromagnetic radiation.

I was also able to discuss multiple aspects of EMP research with Dr. Gary Smith, director of the John Hopkins Applied Physics Laboratory; Janet Danneman, survivability assessment director of the Electromagnetic Division at the White Sands Missile Range; Dr. William Duff, internationally recognized for his work in electromagnetic effects on communication and electrical systems; John Dawson, director of the Electromagnetic Effects Division of the Navy's Air Warfare Center who completed EMP testing at Pax River as early as 1972; Congressman Curt Weldon of Pennsylvania, who served on the House Committee on Homeland Security and was a major voice in Congress regarding the EMP threat; Maty Weinberg with ARRL; Dr. Arthur T. Bradley, electrical engineer with NASA; F. Michael Maloof, former Senior Security analyst for the Secretary of Defense, Rob Hanus, producer of the Preparedness Podcast; Garrett M. Graff, National security journalist; James Wesley Rawles, author of multiple national best-selling books on survival and preparedness; and Tom Brennan with Sol-Ark for the use of their Texas EMP test facilities.

ACKNOWLEDGMENTS

To these experts who provided assistance during my many years of research into EMP and solar storms, plus other government officials and authors who asked to remain anonymous, I want to express my sincere gratitude. Without all their help and direction, this book could not have been written, and finally, thanks to Sharon Yago who provided all of the typing and manuscript preparation.

Introduction

As I was completing the final editing of this book in late March 2020, the Coronavirus, or COVID-19 pandemic was just impacting the United States after starting in China and then raging through Europe and the rest of the world. While this does not directly relate to the subject of this book, it has given everyone in this country a small taste of what an EMP or solar storm will do, which is to cause a major disruption to our way of life and widespread food shortages that could last months if not a year or more.

As I write this, most Americans are currently finding empty store shelves, closed schools and businesses, and the need to self-quarantine at home. They say by summer the virus will have run its course and life will be back on track by the time this book is available. However, imagine what life would be like if these empty store shelves stayed empty indefinitely, while at the same time there was a total loss of all electric power, a major disruption to all forms of travel, closed banks, and failed communication systems. That is what an EMP or solar storm would add to the mix.

While this recent virus pandemic experience was difficult for most families, especially for those who lost their jobs due to closed restaurants and small businesses, those of us in the prepper community were not impacted at all. No, we had not prepared for a fast spreading virus from China raging through the country, but we did prepare for a major disruption to our food supply, loss of electrical power, closed banks, and limitations on travel and communications regardless of the cause.

It really makes no difference if the United States is hit by a pandemic, an EMP attack, or a solar storm, your future survival will

INTRODUCTION

depend on making preparations now. It is my hope this book will show many low-cost ways you can prepare now. After just experiencing the minor disruption COVID-19 has caused, the damaging effects and food shortages caused by an EMP will last years, not weeks, and result in the death of an estimated 200 million people in the United States alone due to starvation and medical complications. So, what is an EMP and how can it cause all this destruction?

By far the most questions I receive when speaking at preparedness conferences these days are – "What exactly is this EMP that everyone is talking about," and "how will it affect me?"

People have a real concern about this thing called "EMP" and they clearly are not finding the answers they want from the main street media. A quick internet search will provide all kinds of articles claiming civilization as we know it will end after an EMP event. An equal number of articles claim the only thing that will happen is a brief power interruption and a few damaged electrical components that can be quickly replaced, and everything will go back to normal.

Some of the articles that discuss an electromagnetic pulse (EMP) event were written by non-technical writers who do not understand the physics behind the phenomenon, and drastically over or underestimate what a real EMP event will do. So, what is true? What is outright fiction? Does anyone really know? If this is a real threat, why isn't the government doing anything about it?

There are real answers to the majority of these EMP questions and yes, there are also still some unknowns. Government and industrial groups have been doing EMP testing since the early 1950s, but most of their research was either "need to know" restricted access or impossible for a non-scientist to understand. At the other extreme, I have found many books on EMP that totally oversimplify the science, yet understanding the science is fundamental to knowing what an EMP can and cannot do to our way of life.

INTRODUCTION

This is where my ten-year quest began, which has taken me all over the country to meet with experts who have worked in this field their entire lives. I have attended national electrical utility conventions and have collected and read a mountain of very dry government and military studies on the subject. I have personally tested all kinds of electronic devices in real EMP test chambers which produced a wide range of results, many of which contradicted the conventional EMP story line.

Scientists working in this field have told me off the record that many of the dire predictions being spread around concerning EMP are just plain wrong. However, some of the same experts have told me a few of these predictions may be right, but for the totally wrong reasons! As they say, just because you are paranoid doesn't mean you are not being followed!

An electromagnetic pulse (EMP) generated by a nuclear detonation at high altitude over the central United States can damage the entire United States' electric grid from coast to coast, which would take years to fully restore. If the power grid is taken down for an extended period of time, store shelves will be empty in days once all deliveries are stopped.

When highways are blocked by abandoned vehicles that have run out of fuel, and gas stations are unable to pump fuel without electric power, all truck deliveries and personal travel will soon end. Any disruption to our electric grid and communication systems will stop all banking and stock transactions, block money transfers and check processing, and shutdown ATM machines and all credit card sales.

Without electricity, refrigerated foods will quickly spoil, and pumps will not be able to pump clean water into our cities or pump the sewage away. Most people can get by temporarily with flashlights and candles for light, and usually find something to eat for a few days. However, without fresh water, people will become very sick in just days after drinking contaminated water and yes, if contaminated surface water runoff is all you have to satisfy thirst, you will drink it.

INTRODUCTION

If in doubt, just read what has happened to the once modern country of Venezuela after their inept government ran their electric grid system into the ground.

During every major power outage in the United States, it took less than a day for all bathrooms in high-rise offices and apartment buildings in affected cities to become totally unusable. Without functioning lift pump stations located at low points throughout every public sewer system, all sewage will immediately back up due to loss of grid power. While many lift stations do have a backup generator, the fuel supply is limited, on the belief service technicians could easily travel and refill the tanks. This will just not be possible after a real EMP event. In addition, how can service and repair personnel get to work sites if the trucks don't run, all gas stations are closed, and there are no parts delivered?

Disease carrying flies and insects will be attracted to the rising mountains of garbage in just days once trash and garbage collection stops, which will add their share of health issues. Hospitals will be overrun by victims of accidents, and increased illnesses spread from eating spoiled foods or drinking contaminated water. Shortages of hospital beds, sterile supplies, and medications will become critical, and the very young and the elderly will be most at risk. Those with long-term medical issues or dependent upon daily injections of refrigerated drugs will go untreated.

The real question is not will all those terrible things really occur, but how long will it take to get things back to normal. Localized power outages in the recent past have clearly shown how dire things can get in a large city in less than a single day when the power is off and there is no water to flush toilets. However, most of these power outages were storm related and usually only paralyzed a single city or perhaps a single state. In each case, help and relief supplies were immediately available from adjoining states not impacted and volunteer relief agencies stood at the ready.

INTRODUCTION

What will it be like if this impacts multiple major cities and multiple states at the same time? What if the power outage and loss of water and sewer services lasts months? Unlike all of these prior utility outages, what if in addition to the power outage there is also a major disruption to all voice and data communication systems, vehicle and public transportation, food delivery, and a total breakdown in civil society?

The most important thing you can do if any part of this country really is hit by an EMP weapon is to immediately recognize what just happened and get out of any large city. This may not be as simple as you might think, since you will not hear an explosion or see a rising mushroom cloud in the distance or buildings blowing over in hurricane force winds so typical in all those doomsday movies. This will not be like the movies. Unlike a temporary power outage, this time in addition to no electric power, most vehicle transportation and all communication systems will be severely disrupted.

If you happen to be in a large city when it strikes, you will have a very limited time before all forms of transportation are hopelessly gridlocked. All traffic and streetlights will be off, and all highway tunnels will be blocked due to lack of operating ventilation fans. If you don't already have a backup plan in place with a safe destination in mind, you will soon be just another victim wondering the streets until you're picked up by security personnel and taken to the closest FEMA camp. Yes, there are hundreds of empty FEMA camps surrounded by high security fencing and cameras scattered around this country just waiting for this type of major disruption to civil society to occur, regardless of the cause.

An EMP is not the only risk for a massive grid down event. The sun is also capable of causing extensive damage to our electric grid when the earth passes through a plasma discharge ejected from the sun's surface. This has happened multiple times before in our past which will be discussed in far more detail in chapter 4. In reality, it's not the actual EMP or solar storm we need to fear, it's the resulting

INTRODUCTION

failure of our electric grid and all communication systems that will cause the real damage. The loss of this infrastructure will require years to rebuild and will cause the death of millions of people due to starvation, disease, and lack of medical care.

It is my hope this book will provide the reader not only with a better understanding of these risks, but also provide ways you and your family can continue to live comfortably until our electrical infrastructure and civil society have been restored.

This book is divided into three major parts: Chapters 1 through 4 provide a more detailed understanding of an EMP and solar storm. Chapters 5 through 13 describe how an EMP will impact each of our major utilities and communication systems. In the remaining chapters I recommend what each of us can do to be better prepared, including simple ways to protect our own electronic devices. There are extensive footnotes to document every fact and all statistical data I have included. I strongly recommend you take each chapter in order, as each builds upon the material addressed in each prior chapter and remember, knowledge is also power.

CHAPTER 1

Understanding Electromagnetic Disturbances

I hope you will not just read the title of this chapter and skip the material because it sounds too technical. If you want to be able to separate EMP truth from hype, we really do need to first review the basics.

The outer space our earth passes through as we circle around the sun every 365¼ days each year is constantly being bombarded by dangerous high-energy X-rays, gamma rays, cosmic rays, and multiple forms of electromagnetic radiation. Fortunately, our earth is surrounded by an outer protective layer called the "ionosphere" which begins around 40-miles up and extends to 250-miles above the earth's surface. Since this ionized layer expands as it absorbs this cosmic and solar radiation, we sometimes consider 600-miles up as its highest possible upper limit.

At this high-altitude there is basically no atmosphere, and the few air molecules that rise up to this level are quickly impacted by all this cosmic radiation. These impacts "knock off" an outer electron of these air molecules, which produces a positively charged particle

ELECTROMAGNETIC DISTURBANCES

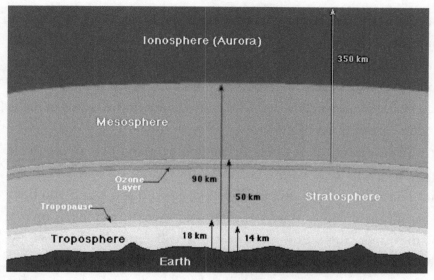

Fig. 1-1. Protective ionosphere surrounding earth.

called an ion. In other words, this outer ionosphere is basically a dense cloud of positively charged particles blanketing the earth and providing a protective layer that absorbs these damaging X-rays and gamma rays and keeps them from reaching the earth's surface. Without this protective barrier, this high level of electromagnetic energy would kill every living thing on the earth.

The free electrons given off when air molecules are bombarded by all this cosmic energy are attracted to the earth's magnetic lines of force and follow these "paths" down to the surface. The thickness or density of this protective layer changes based on how much radiation it absorbs. Since a large percentage of this electromagnetic radiation reaching the earth is being ejected from the sun, the ionosphere covering the daylight side of the earth will be denser and raise to a higher altitude than the ionosphere over the dark side of the earth.

Since the earth's axis of rotation is actually tilted by almost twenty-four degrees, this ionosphere density over the earth's pole that is tilted away from the sun will have a correspondingly lower density. Due to this tilted axis, the North Pole will tilt away and

ELECTROMAGNETIC DISTURBANCES

northern latitudes will face more towards the sun during our summer months, which will provide a greater solar exposure to the United States during half of the year. Since certain upper layers of the ionosphere are excellent "reflectors" of radio waves, this is why you can hear very distant AM and ham radio stations more clearly at night and at certain times of the year.

The earth's core is primarily molten iron and like all metals, it has a magnetic field with a north (N) and south (S) magnetic pole. Just like the magnets you may have played with as a kid, if you hold a bar magnetic under a sheet of paper and sprinkle iron filings on top of the paper, these tiny metal particles will immediately move by themselves into curved lines spreading out at the center and curving back towards the magnetic "N" and "S" ends of the bar magnet. Although we cannot actually feel or see these invisible magnetic lines of force, these tiny metal-particles have no problem immediately finding them and lining up without any assistance.

The earth has the same type of magnetic lines of force, only these lines are much larger and stronger, and are spaced out all around the circumference of the earth and originating out from each pole. The magnetic N and S poles generated by the earth's molten iron core do not line up exactly with the earth's geographic North Pole and South Pole, but for all practical purposes they are in the same general area. The "N" end of a compass needle will always point to the earth's magnetic South Pole, which is actually located at the geographic North Pole. Confusing!

ELECTROMAGNETIC DISTURBANCES

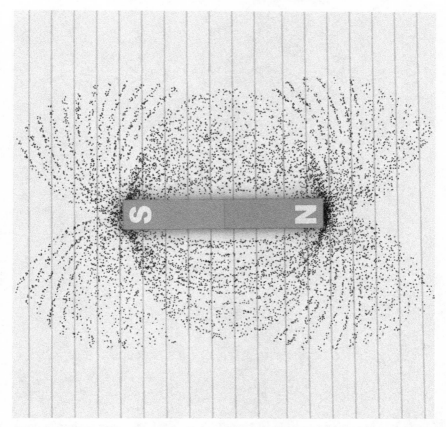

Fig. 1-2. Iron filings quickly line up to show the location of magnetic lines around a bar magnet.

It should be noted that the magnetic north and south poles of the earth slowly shift away from the geographic north and south poles over time, as the earth's molten inner iron core changes magnetic orientation. There have been multiple times in the earth's prehistory when the magnetic N and S poles actually reversed with each other. However, the earth's tilt and rotation did not change, and the earth did *not* actually "flip upside down" when this happened!

ELECTROMAGNETIC DISTURBANCES

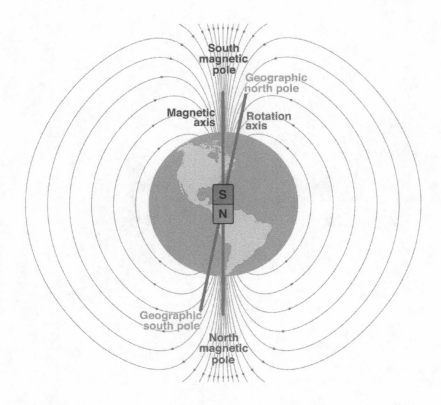

Fig. 1-3. Magnetic lines of force surround the earth, but the magnetic poles do not line up with the geographic poles.

This unseen magnetic field and the upper ionosphere work together to provide this protective barrier around the earth to deflect away or absorb the harmful electromagnetic radiation constantly bombarding our planet. However, any disruption to this magnetic field can temporarily alter its ability to block this harmful radiation. Understanding how the magnetic lines of force and the ionosphere surround and protect our earth makes it easier to understand how any disruption to this magnetic field caused by an electromagnetic pulse (EMP) or solar storm can severely impact our electric grid.

ELECTROMAGNETIC DISTURBANCES

If your science teacher ever demonstrated electricity by moving a bar magnet back and forth across a wire, you saw the needle of a connected voltmeter indicate that a tiny electrical current was being generated. Now imagine how much electrical energy hundreds or even thousands of miles of cross-country electric lines will generate when the earth's magnetic lines of force are shoved back and forth across the entire length of these long wires. One of the primary effects of an EMP or solar storm event is how they shove aside the earth's magnetic lines of force. This effect will be discussed in more detail in chapter 3.

Extremely high voltages and thousands of amps of DC current can be induced into long cross-country power lines when the earth's normally stationary magnetic lines of force are "shoved" back and forth across these long wires. These magnetically generated currents and voltages then travel down these cross-country wires and into the sub-stations and large high-voltage AC transformers located throughout the grid system. These grid transformers cannot handle ground-induced DC currents which are imposed on the neutral conductors of these transformers by an EMP, causing them to overheat and self-destruct.

During the nuclear bomb testing that took place in the late 1950s and early 1960s by both the United States and Russia, extremely high DC voltages were generated in long utility lines. This was how this effect was first discovered, although scientists had long predicted this might be one of the side effects of a nuclear detonation. Surprisingly, it made little difference if the wires were stretched on above ground electric poles, or buried deep below ground level, high DC voltages were still induced into the wires.

The next chapter will provide a better explanation of the multiple ways the earth's magnetic field can be disrupted, which can not only damage the long cross-country wires and large transformers

ELECTROMAGNETIC DISTURBANCES

in the electric grid, but also destroy many devices electrically connected to them.

We have been discussing an EMP and its effect on the ionosphere which then affects the electric grid, but what is it? Basically, electromagnetic waves are just a form of energy having both electrical properties and magnetic properties, with each component acting at a 90-degree angle from each other, in reference to the direction this "wave" travels through space. Originally thought to be two separate things, we now know these two different properties of an electromagnetic wave are each actually half of the same thing.

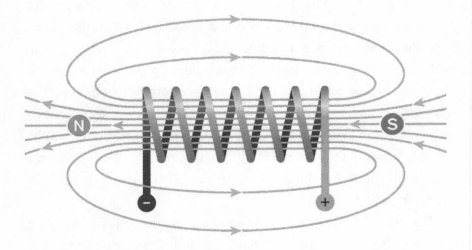

Fig. 1-4. A wire carrying a current has both electrical properties and magnetic properties.

As introduced earlier, when we move a magnet across a wire it will produce electricity, but if we reverse the process and run electricity down the same wire, it will create a magnetic field around the wire that you can actually observe with a small compass. Electromagnetic waves can have many different frequencies (cycles per second) just like plucking the string on a guitar. The shorter or tighter the string, the higher the frequency of the wave produced.

ELECTROMAGNETIC DISTURBANCES

The fewer number of up and down cycles a wave makes per second, the longer the length from start to finish of one complete cycle of the wave frequency, which is measured in meters not feet, (a meter is slightly longer than a yard). The frequency of the energy waves striking the earth can be anything from a few hundred to a gigahertz or more cycles per second. Throughout this book the term kilohertz (kHz) is defined as a frequency wave of 1,000 cycles per second. A megahertz (MHz) is 1,000,000 cycles per second, and a gigahertz (GHz) is 1,000,000,000 cycles per second.

Lower-frequency electromagnetic waves have a longer wavelength and are very similar to the type of electromagnetic waves generally used to broadcast AM and shortwave radio transmissions. As the frequency of a given wave increases, its wavelength becomes shorter and shorter, with microwaves, then visible light, then X-rays, followed by gamma rays representing the longest to shortest wavelengths. However, our eyes are only able to see the small visible light spectrum of this wide range of electromagnetic frequencies.

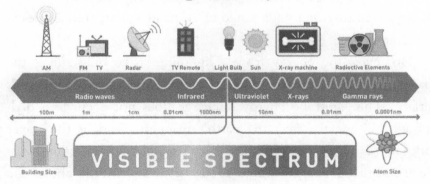

Fig. 1-5. The electromagnetic spectrum from very-low to very-high including visible light.

ELECTROMAGNETIC DISTURBANCES

An electromagnetic energy field can consist of any number of electromagnetic waves having many different frequencies across many different wavelengths, depending on the power and type of energy source that created them. Our bodies are constantly being bombarded by all kinds of electromagnetic energy in the form of radio waves, television waves, wireless internet, shortwave radio transmissions, cell phone transmissions, and even a few extremely high frequency X-rays and gamma rays from outer space that leaked past our protective ionosphere. Thankfully these waves normally pass right through us without any harm due to their extremely low power levels. However, if we could boost the power of these energy waves millions of times, they would be extremely damaging. It's the increased power of the source that produces an energy wave that causes it to be so destructive.

Another form of electromagnetic energy is lightning. If you have ever listened to an AM radio station during a nearby lightning storm, you will hear static from the electromagnetic discharge between the clouds and the earth, regardless of which channel or frequency your radio was tuned. A lightning strike generates a very high voltage energy consisting of a wide range of electromagnetic frequencies at the same time.

However, the energy from a lightning strike only affects the space around and under the clouds covering a small area of the earth's surface. This lightning energy discharge does not occur at a high enough altitude to affect the earth's magnetic field or the upper ionosphere. To really impact the earth's magnetic field requires the electromagnetic energy source to occur at an altitude above the ionosphere.

While a nuclear detonation near ground level will generate lots of electromagnetic energy, the damaging effects to any electrical wires or equipment is limited to a small radius around the blast site. However, if a nuclear detonation occurs high above our upper atmosphere, both the earth's magnetic lines of force and the protective ionosphere would be severely impacted by this electromagnetic "shockwave" produced by any high-altitude nuclear detonation.

I will be discussing various intensities of extremely strong magnetic fields that are generated by this electromagnetic pulse in the following chapters. A magnetic field is the direct result of an electrical flow. In this case the flow of electrons traveling down to the earth while following the earth's magnetic field, after they are created by a high-altitude nuclear explosion. The earth's magnetic lines spread apart the closer you get to the earth's surface, as these serve as "highways" for this flow of charged electrons to reach the earth's surface. This increasing distance between these magnetic lines and the electron flow following them causes an increase in the strength of the electromagnetic fields being generated. In addition, an energy field of opposite charge will develop in the earth immediately below the electron flow above to "drain off" or ground the charge.

The measurement of electromagnetic fields is based on the distance between separately charged surfaces, or points in space, measured in meters. The actual term for describing the strength of this field is "kV/m" which is 1,000 volts of electrical energy potential per meter of separation of the two surfaces. For example, testing has shown a magnetic field of 10 kV/m is more than enough energy to damage any unshielded electronic devices within this field.

Most design specifications for the magnetic shielding of military communication systems use a value of 50 kV/m of field strength, which was the maximum field ever measured during upper atmosphere nuclear detonations back in the early 1960s. In addition, scientists have learned that 90 percent of this electromagnetic field consisted of frequencies in the upper 100 kHz to 1 GHz range.[1] A second term used to measure the magnetic field induced by a wire or coil of wire is "A-turns/m." This describes a magnetic field around a coil of wire having a given number of amps passing through each turn of wire in a coil, as measured by the length of the coil measured in meters.

For reference, a magnetic field around a wire coil having a value of 100 A-turns/m is approximately the same magnetic intensity as the previously mentioned 50 kV/m standard used to measure the intensity between two plates or points in space. These values and their upper frequency range are very typical for the electromagnetic fields generated by a nuclear EMP.

It should be noted that recent developments of nuclear devices specifically designed to generate an electromagnetic pulse as the main purpose, and not as a byproduct, have achieved electromagnetic intensities reaching 100 kV/m. This is double the 50 kV/m standard currently being used to design EMP protection for all military communication systems. A very sobering thought for our military planners!

Protective shielding may block most, but cannot block all, of the higher frequencies contained in an EMP energy field in its 100 kHz to 1 GHz range. For example, both a taped and sealed trash can, ammo can, and discarded microwave oven can easily protect against EMP energy in the lower 100 kHz. However, as frequencies approach 1 GHz, their shielding effectiveness drops by over half.[2] Since an EMP will include varying intensities of all frequencies between these upper and lower limits, obviously additional levels of shielding will need to be added to provide 100 percent protection against all possible EMP generated frequencies.

CHAPTER 2

What Is an EMP?

EMP was discovered as a by-product of the first high-altitude detonation by the United States of an atomic bomb in 1962 known as Starfish Prime. The United States detonated this 1.4 megaton nuclear bomb approximately 250-miles above Johnson Island, which is a small uninhabited island in the Pacific. Years earlier scientists predicted there could be some type of energy shockwave created when a nuclear bomb was exploded at a high altitude, but until this test all nuclear testing was done at or near ground level. This was the first detonation of a nuclear bomb outside the earth's atmosphere.

There has been much misinformation about this nuclear blast that caused some disruption to the electrical grid and telecommunication systems in Hawaii almost nine hundred miles away. Follow-up studies indicated that out of over ten thousand streetlights, a large percentage went dark due to an estimated ten to thirty fuses being blown by the high-voltage surge induced into the streetlight power wires, but only a few bulbs actually burned out. There was no damage reported to any grid power lines, circuit breakers, or transformers.

There was also no loss of telephone or radio service in Hawaii, although damage was reported to a microwave repeater station and several radio transmitter power supplies. The phone system and all radio and television stations continued to operate. The Hawaiian Islands were still a long way from the blast and being islands, their electric lines were only a few miles in length. However, the nuclear

WHAT IS AN EMP?

detonation did create a temporary radiation belt above the earth that temporally blocked out radio reception and produced illuminated night skies thousands of miles away.[1]

A series of seven high-altitude nuclear bomb tests of varying sizes of warheads were conducted by the Soviet Union later in 1962 which were known simply as Test 184 and these did cause significant damage to civilian electrical systems since they were detonated high above land and not an isolated island in the middle of the ocean. The first three hundred-kiloton nuclear bomb was detonated 180-miles above Kazakhstan which is a large industrial region larger than Western Europe. The smallest warhead tested resulted in an electromagnetic shockwave with a horizontal radius of approximately twelve hundred miles and took down their entire electric grid. This test damaged a six-hundred-mile-long section of underground shielded power line buried ten-feet deep, which started a fire in a power plant connected to one end. Unlike the small island of Hawaii with relatively short power lines, the longer cross-country power lines provide a more damaging "coupling" to the EMP pulse.

Test 184 also caused damage to above ground power lines, phone lines, and damage to the internal copper coil windings in several large diesel generators. These generators later failed due to the breakdown of the wire insulation in the generator's coil, which was not discovered until some weeks after the blast. Since most of the nuclear warheads used in this test series were relatively small, in order to cause this much EMP damage it appears Russia was developing a super-EMP weapon. It is theorized at least some of these nuclear warheads were modified to increase the production of gamma rays, which in turn produced a much stronger E1 shockwave.[2]

It has been more than sixty years since these high-altitude nuclear tests were carried out. No doubt any country having nuclear capability, including the United States, have been developing even more powerful EMP weapons with smaller and smaller nuclear

WHAT IS AN EMP?

triggering devises. In other words, to think this technology will not be part of a future attack on the United States is shear fantasy.

In chapter 3 I address how a high-altitude nuclear explosion generates a large EMP when this energy interacts with the earth's magnetic field. Since all above ground nuclear testing ended in the early 1960s, we really have very little research on the effects this can cause to more modern types of electrical equipment.

The electronics of the early 1960s was hard wired and vacuum tube based, which could withstand induced voltages and currents thousands of times higher than today's tiny microelectronic devices. While our electrical transmission lines may withstand the effects of an EMP attack, it is doubtful most of the high voltage transformers and electronic devices connected to the ends of these powerlines will survive.

Since actual testing of today's microelectronic devices to withstand the effects of a real high-altitude nuclear detonation is no longer possible due to international test ban treaties, numerous testing chambers have been built which can simulate (on a small scale) a high-energy solar storm and an electromagnetic pulse (EMP) event. In many cases, even if the electronic device was not physically damaged, the EMP can still cause data errors on magnetic memory devices, data transfer errors on Ethernet cables and calibration problems with electronic sensors.

This data error issue is troubling since all power plants have multiple large turbines and pumps that require a very orderly and controlled plant startup and shutdown process. With electric motors the size of a pickup truck you don't just turn them on and off with a wall switch. The EMP energy entering a power plant through the power lines can damage the computers controlling these large motors, resulting in major damage to these large pumps and fans due to out-of-control speed regulation or improper shutdown procedures. Similar concerns exist with long telecommunication lines. Any communication lines that are fiber optic based instead of copper will not absorb and carry a high-voltage pulse down the cable, but the sending and receiving units at each end that utilize microelectronics will be damaged.

Electromagnetic test chambers have shown it takes less than 10 watts of power to destroy today's extremely small microchips and integrated circuits (IC) if the pulse peaks in only a few nanoseconds (10^{-9} sec). However, the more distant the device is from the source of the pulse, the lower the strength of the energy and the less the damaging effects. This would indicate that even unprotected electronic and computer devices may experience little or no damage if the distance from a nuclear detonation is increased significantly. This assumes these devices are not connected to very long power lines or tall antennas, which can amplify and inject the "collected" electromagnetic energy from a more distant EMP into these more distant electronic devices.

While the news media has convinced us that only a nuclear weapon exploded high above the earth's surface can produce an EMP, this is not the case. Technically, this more common form of EMP is called HEMP, which indicates it is caused by a high-altitude nuclear detonation. However, many countries, including the United States, have EMP generating weapons, that do not require a nuclear detonation at all to generate EMP energy. While these non-nuclear EMP weapons can only impact a small area like a military base or small city, they can still generate EMP energy levels more than double that produced by a high-altitude nuclear bomb, which requires a much higher level of EMP protection.

An EMP weapon can easily fit inside a military drone or in the back of a pickup truck, but obviously will have a limited range. EMP weapons this size are more directional in their discharge. This can make them more suitable to destroy the electronics and computer systems in a large building from a nearby vehicle in the parking lot,

WHAT IS AN EMP?

or perhaps a city block of buildings if the device is somewhat larger. However, we are still talking about a fairly short range for these non-nuclear EMP weapons. There will be no damage to the structures or people inside, but the damage to the computers, phone systems, internet routers, and electrical equipment will be fairly extensive, except for any shielded devices connected with shielded cables.

An EMP weapon can even be made small enough to fit inside a briefcase, but still powerful enough to destroy the electronics in a nearby computer system or vehicle. It is common for larger police departments to end a dangerous car chase through crowded city streets by pulling behind the fleeing vehicle and firing an EMP "gun" into the vehicle just ahead. This will cause major damage to the vehicle's ignition system, dashboard electronics, and in extreme cases even cause the brakes to lock up or the airbags to deploy.

In chapter 3 I will discuss in much more detail the basics of an EMP event and chapter 6 will go into much more detail regarding how a real EMP event will affect most of today's modern vehicles full of computerized devices.

CHAPTER 3

EMP Basics

An Electromagnetic Pulse (EMP)is basically an energy shockwave generated by a nuclear bomb exploded far above the earth's surface. It is totally silent, since the nuclear detonation that produced it was several hundred miles above the earth where there is no atmosphere to carry the sound. Since an EMP is electromagnetic energy, it does not affect people or animals, and there is no fireball type explosion or pressure wave at ground level, so no buildings or other structures will be damaged.

In fact, at first you probably would not know an EMP event actually occurred. Initially it will appear to be just another power outage, except in addition to all of your electric lights and appliances not working, there will be a high probability many vehicles and all radio and phone communication will experience problems. We know firsthand how a nearby lightning strike can damage household appliances and electronic devices, as this electrical discharge between earth and sky can generate a very high-voltage "spike" in the power lines supplying electricity to your home.

From a technical standpoint, one difference between EMP and lightning is the electrical discharge from a lightning strike takes five microseconds to reach its peak level, while an EMP energy wave takes only a few nanoseconds to reach its peak voltage. So, what's a nanosecond, and why is that important? It takes very small slices of a single second to describe an EMP event. Throughout this book a millisecond is defined as one-thousandth of a second, a microsecond as one-millionth (10^{-6}) of a second, and a nanosecond as one-billionth (10^{-9}) of a second. EMP energy reaches its peak one thousand times faster than lightning peaks. We are talking about dividing a single second into 1,000,000,000 tiny parts in order to measure the timing of an EMP!

Perhaps an easy way to visualize why the shorter ramp-up time is important when considering an EMP event is the impact of a car crash against a solid wall. If we have two identical cars having the same weight and size and drive the first car into a solid wall at five miles per hour and drive the second car at sixty miles per hour, the car having the fastest rate of speed will have the highest energy potential and cause the most damage. Although this may not be an apples-to-apples comparison, it at least helps explain how any electromagnetic shockwave, regardless how it is generated, will result in increased damage to electrical devices and be harder to block if it has the power to compress the time needed to reach its peak power level by a factor of over a thousand times faster than a lightning strike.

The use of lightning arrestors or surge suppressors to block the high-voltage surge imposed on nearby electric lines by a lightning strike will not offer any protection from the initial phase of an EMP pulse, as these voltage limiting devices just cannot react fast enough. Surge suppressors and lightning arrestors designed to filter or block the high-voltage spikes from lightning reaching sensitive electronics downstream cannot open up fast enough or shunt this high-voltage peak to ground if this energy pulse is the result of an EMP event.

For example, cell phone and television towers are struck by lightning on a fairly regular basis, yet by using lightning arrestors and good grounding methods these lightning strikes cause little or no damage to the computers and transmitters connected to these towers. Unfortunately, these protective devices will not even "see," let alone react fast enough to block an energy spike traveling over a thousand times faster than the typical voltage spike induced by a lightning storm. A voltage spike from an EMP event will pass straight through these safety devices as if they were not even there. While the primary difference between EMP and lightning is how fast they reach their maximum peak level, there is also another major difference between lightning and EMP.

While lightning is basically one large electrical discharge with everything occurring as a single event, the energy discharge from an EMP actually consists of three separate segments called E1, E2, and E3, and each segment consists of different frequencies and take different lengths of time to ramp up to their peak energy levels.

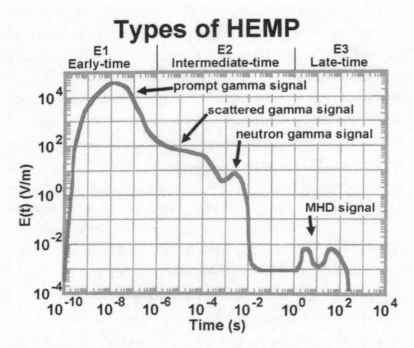

Fig. 3-1. Graph showing three phases of an EMP verses time each occurs.

The E1 signal will reach the earth's surface almost instantly, at full power and consist of very high frequencies in the range of 100 kHz to 1 GHz. In fact, 90 percent of the energy contained in this E1 phase of an EMP is in this higher frequency range.[1] This very high frequency and fast peak time means the initial E1 phase will travel through walls, roofs, or just about anything except solid metal, so E1 energy does not need a long wire to carry its damaging effect into any electrical or electronic devices. This is the main reason why these "black box" EMP protection devices being advertised to totally protect your electrical appliances from an EMP cannot do what is being claimed. Notice in the EMP graph how the section of the curve representing the initial E1 phase is going almost straight up to its peak almost instantly.

The E2 portion of an EMP event is a little slower to ramp up to its peak level and consists of the middle frequencies range of 1 kHz to 1 MHz and with less power than E1. Since the E2 part of EMP event takes almost one thousand times longer to reach its peak level, it is less destructive. In fact, the slower E2 middle phase an EMP event is very similar to a lightning strike and a high-quality lightning arrestor can block E2 energy from passing through into house wiring from the electric grid and destroying electrical devices.

Unfortunately, since the slower E2 part of an EMP event lags behind the extremely fast E1 energy, there is a high probability most lightning protection devices will be damaged by the E1 energy the instant it passes through, so these devices may no longer function properly to block the thousand times slower E2 phase.

We learned in chapter 1 how anything that can affect the earth's magnetic lines of force will induce high-voltages into long utility wires. Each mile adds to this radiated energy being collected along the entire length of these wires, and this induced voltage is routed into downstream transformers and switchgear causing major damage.

A nuclear bomb detonated 250 miles above the earth will produce extremely high frequency gamma rays and X-rays. During the explosion, these gamma rays and X-rays expand out in all directions, but those heading towards the earth strike air molecules in our upper atmosphere causing an electron from each air module to break out of its molecular orbit and continue towards the earth with the energy gained from its impact with the gamma rays.

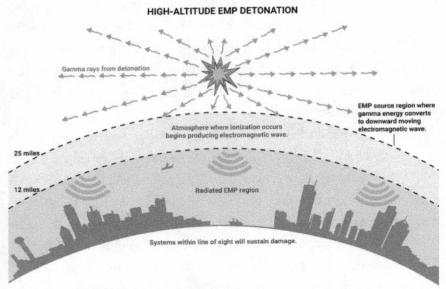

Fig. 3-2. Energy produced from high altitude nuclear detonation.

This is similar to playing pool, when the fast-moving cue ball stops dead in its tracks when it impacts a stationary ball and transfers its energy and motion into this impacted ball. This resulting flow of negatively charged electrons acting together create a very high DC voltage field moving down towards the earth. The physical process of a high-energy gamma rays knocking off an electron from an atom in an air molecule, which then heads off in the same direction as the gamma ray was traveling is referred to today as the Compton Effect or Compton Scattering. The resulting free electron is called a Compton Recoil Electron, named in honor of Dr. Arthur H.

EMP BASICS

Compton who discovered this effect in 1923 for which he was awarded a Nobel Prize.

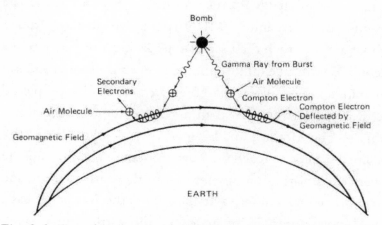

Fig. 3-3. Free electrons as a result of gamma rays striking upper atmosphere air molecules follow earth's magnetic lines down.

These negatively charged electrons follow the earth's magnetic lines as if they were wires, which causes this very high-frequency E1 and middle frequency E2 energy to travel down to the earth's surface. Here this electron flow will be "collected" by any wires, pipes, metal roofing, or any metal object which act as an "antenna" causing the metal objects to be electrically charged.

The E3 electromagnetic force produced by the explosion of a nuclear bomb at high altitude acts totally differently and is even produced differently than the much faster and higher frequency E1 and E2 energy waves. While E1 arrives almost instantly, and the slower E2 wave takes a fraction of a second longer to reach its peak, the E3 energy that follows both E1 and E2 is totally different. The E3 part of an EMP event has a much higher magnetic component which causes the earth's magnetic lines of force to actually be "pushed aside" as it passes. This is the same type of disturbance to the earth's magnetic field a solar storm generates, which I address in chapter 4.

The E3 phase of an EMP event is often referred to as the late time "magnetohydrodynamic" (MHD) pulse. Say that five times! Although E1 and E2 reach their separate peaks in a tiny fraction of a single second, E3 can take up to a minute to even get started and can last many minutes. This allows the E3 energy a much longer period to affect the long power lines than a voltage surge lasting just a fraction of a second. Both E1 and E2 energy waves follow the earth's magnetic lines of force to reach the earth from the upper ionosphere with a "vertical" polarization in reference to the earth's surface and similar to radio and television waves. However, the E3 energy has a "horizontal" polarization in reference to the earth's surface and consists of very low frequencies.

Immediately after the high-altitude nuclear explosion occurs and right after the ejection of the gamma rays and X-rays that generate the E1 and E2 electromagnetic energy waves, there will be extremely high temperature bomb debris and additional high temperature ions blasting out and down from the nuclear detonation. In fact, while the gamma radiation which cause the generation of both E1 and represent less than 0.10 percent of the total explosive energy of the nuclear bomb blast, over 25 percent of the blast's energy is in the debris field exploding out, followed by the remaining 75 percent of the bomb's energy in X-rays. Together they represent the bulk of the energy driving the generation of the E3 phase of an EMP, which is far greater than the energy contained in both E1 and E2.

The expanding nuclear fireball forms a "magnetic bubble" in the first few seconds after the initial blast, quickly followed by the arrival of the extremely hot debris field. Over the next few minutes this area or "patch" of the upper ionosphere directly below the blast is superheated by the shockwave of heat and electrically charged ions "raining down" which causes an electrically charged magnetic "bubble" to expand. Thus producing more buoyancy in this patch of the ionosphere causing it to rise at supersonic speeds from the lower to upper level of the ionosphere. As this electrically charged layer rises, it crosses and distorts the fairly stable magnetic lines surrounding the earth.

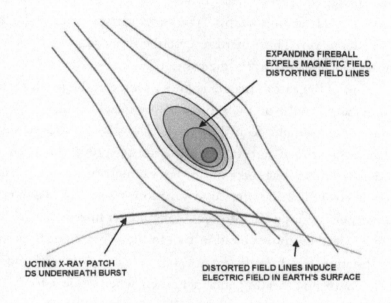

Fig. 3-4. As it expands out, the nuclear fireball distorts the earth's magnetic field.

This interaction generates extremely large current flows within the rising super-heated patch which moves in a north and south rotation.[2] This in turn produces a mirror image current flow at the earth's surface directly below this collapsing magnetic bubble inside the rising patch of ionosphere. Think of playing with two bar magnets as a kid. You wave one magnet above the other bar magnetic laying on the table and it moves to line up with the correct magnetic field.

This coupled energy flow in the earth acts at right angles, or east to west and west to east as a *horizontal* electromagnetic field. It's this horizontal acting field that gives it the ability to "shove" the north-south magnetic lines of the earth from side to side. In other words, unlike the initial E1 and E2 phase of higher-frequency energy, the much lower frequency and slower E3 energy produces huge magnetic fields acting "across" the earth surface, much like high winds in a storm. This is hard to describe in words, but I hope this explanation at least conveys the basics.

Think of the earth's magnetic lines of force as the parallel strings on a guitar lying flat on the ground. Now imagine what the earth's magnetic lines would do if they were "plucked" or shoved side to side in a horizontal direction while crossing power lines that are hundreds of miles long. Like our earlier example of waving a magnet across a small electric wire, large very low frequency DC voltages and currents are induced into these long power lines, which flow in the wires leading into substation transformers located at both ends. All substation transformers are AC only electrical devices and will suffer catastrophic heating and meltdown when subject to the DC currents imposed on the neutral conductions by an EMP.

Every day the ionosphere above the region of the United States is heated and expands each afternoon as more ions are produced during the daylight hours as the sun's energy is absorbed. However, a nuclear bomb would produce less heave or lift of a patch of ionosphere and less E3 energy if the ionosphere is already heated and charged. This would tend to favor a morning bomb detonation if the goal is to generate the most E3 energy destruction. In addition, there will be a period that could last several days in which the ionosphere is totally saturated after an initial EMP generating detonation, so if there is a second nuclear detonation, it will be delayed until the ionosphere re-stabilizers to maximize the damage. This would indicate it might be wise to delay the replacement of any damaged electronic equipment until the military can verify and stop any attempt to cause a second high-altitude detonation.

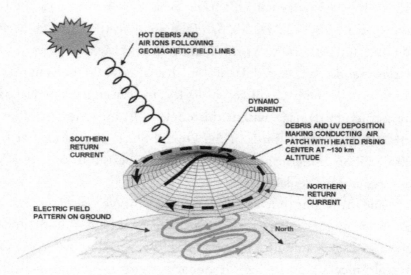

Fig. 3-5. Upper atmosphere "heave" caused by a nuclear detonation generates the E3 phase of an EMP event.

Tests have shown the most disturbance to the earth's magnetic lines occurs at the outer edges of this rising patch or heave area and produces the most magnetic field lines facing north due to the earth's magnetic field sloping down as it nears the poles, just like the bar magnetic example discussed in chapter 1. To maximize the E3 energy which will create the most damage to the electrical infrastructure of the United States, the center of the heave or patch area of the ionosphere will need to be located south of the United States, or centered over Mexico.[3] This would be easy to accomplish with a near vertical launch from a container ship located in the southwest portion of the Gulf of Mexico. However, that may not be the ideal burst location to maximize the damaging effects of E1 and

E2 to our country, which would favor a more central United States location.

A single high-altitude EMP, or HEMP, with a detonation 250-miles above Omaha will still have over 50 percent of its total E1 energy reaching Los Angeles, San Francisco, Seattle, Atlanta, Miami, Washington DC, New York, Winnipeg, and Mexico City. Within a one-hundred-mile radius of Omaha, almost 90 percent of the resulting E1 energy, will reach 40 kV/m, causing major electrical infrastructure damage within this circle.[4] Within one year after an HEMP attack, two-thirds of the United States population, or over 200 million Americans, would probably perish from starvation, disease, and the social collapse.[5]

The damaging effects this magnetic "heave" and superheated patch of the ionosphere that is generated by a high-altitude nuclear bomb is also created by a solar storm. However, the solar storm does not produce E1 or E2 high frequency energy which are generated by a nuclear EMP. Chapter 4 will address the damage the E3 late-time electromagnetic energy can do to the nation's electric grid. This means a solar storm will cause the same major damage to the nations' electric grid and most high-voltage transformers, but will not affect cars, or damage electronic devices.

CHAPTER 4

Don't Forget the Sun

As stated in chapter 1, the earth is constantly being bombarded with all kinds of electromagnetic energy from the sun and outer space during every second of every day. The majority of the sun's electromagnetic energy is in the form of infrared, visible, and ultraviolent-light, which provide all of the earth's natural illumination and all of its warm-up again after each ice age, although this historical fact continues to allude most global-warming advocates!

Sometimes dark spots will appear on the sun's surface which are an indication of a major electromagnetic disturbance to the earth is about to occur. Sometimes these sunspots will produce a violent burst of electromagnetic energy called a Coronal Mass Ejection (CME). These bursts of energy shoot out from the sun's surface in the form of high-energy X-rays and gamma rays traveling near the speed of light and reaching the earth in only a few minutes, followed by a "cloud" of energy-charged particles or plasma which travels slower and can take a day or more to reach us.

There are also times when these clouds of plasma shooting out from the sun's surface are much weaker and spread out in a more diffused field of energy called a solar storm. While a solar storm contains less electromagnetic energy than a CME, they still can have a detrimental effect on our electric grid and communication systems for several days if this large energy field crosses the earth's path. Just remember, a solar storm is not as powerful and will not cause as much damage as a CME, but most articles use both terms interchangeably.

DON'T FORGET THE SUN

The dark sunspots which produce CME's and solar storms typically occur on an eleven-year cycle when a maximum number of dark areas will be visible on the sun's surface. Sunspots can occur anywhere on the sun's spherical surface and can shoot out into space in any direction at speeds of millions of miles per hour. Occasionally one of these major streams of electromagnetic energy will shoot out like water from a fire hose and cross directly in front of the path of the earth as we revolve around the sun. If the timing is just right, this path of high-energy plasma will actually strike the side of the earth facing the sun right when we cross this path during our orbit around the sun.

The only good news is it takes several minutes for the bombardment of the higher-energy X-rays and gamma-rays to reach the earth's path, and up to eighteen hours or more for the plasma cloud that follows to travel the same distance. This allows astronomers, and the satellites that constantly monitor the sun, to provide an early warning to announce the earth is about to be impacted.

There have been multiple years with no sunspot activity at all. The last time this occurred was called the Maunder Minimum, which lasted from 1645 until 1715 and caused the famous Little Ice Age all over the world. Rivers and large lakes that had always been ice-free during every winter totally froze over, and many areas of Europe went through multiple summers when winter snows never completely disappeared off farmers' fields.

Fortunately, the sun finally provided the warmup needed to bring us back to a more temperate climate. Since this major warmup occurred in the early 1700s, I am reasonably sure this global warming was not caused by increased CO_2 levels or the industrial age! As noted earlier, it is extremely rare when a CME crosses the earth's path right when the earth reaches the same point in space and pass through this electromagnetic energy cloud, but this actually did occur in 1859.

At this point in our history, Edison was still twenty years away from perfecting the light bulb. Nikola Tesla was still twenty-five years away from inventing AC electricity, which would later make our electric grid possible. Marconi was thirty years away from inventing radio transmission, so homes still did not have any electrical appliances or radios. However, Samuel Morse had invented the telegraph and had established telegraph lines between all major cities by 1844, so in 1859 the United States and Europe were crisscrossed by a large network of telegraph wires. Since homes were still illuminated by kerosene or gas lamps, in 1859 the telegraph offices contained the only electrical devices in the world at the time.

No electric lightbulbs, microwave ovens, cell phones, computers, internet routers, credit card machines, DVD players, video games, air-conditioners, satellite receivers, dishwashers, toasters, radios, washing machines, elevators, refrigerators, calculators, clothes dryers, electric motors, radio transmitters, flat-screen TVs, laser printers, water-pumps, transformers, or vehicle electrical systems suffered any damage, since not a single one of these devices existed in 1859!

Our country's first experience with the destructive effects of a CME occurred on September 1 and 2, 1859 when the night skies all over the world started to glow. Reports of shimmering colors and cloud-like glowing effects in the sky, similar in appearance to the northern lights (aurora borealis), which can be observed today in northern latitudes, but much, much brighter. News accounts of the event describe being able to see perfectly well in the dead of night, and sleeping birds were awaken and flying around thinking it was morning.

While I am sure at the time this was beautiful to watch, this solar storm caused major damage to the only electrical equipment in use in 1859 – the telegraph system. The same energy that had just been ejected from the sun the day before, was now illuminating the night sky. This burst of electromagnetic energy induced high-voltage flows in thousands of miles of telegraph lines from one end of the country to the other in both the United States and Europe. These wires carried this high-voltage electrical energy down the lines and into the hundreds of telegraph offices located at the ends of each line segment. Some telegraph operators were shocked, and some telegraph offices actually caught fire from the resulting sparks emitted from the code sending and receiving devices.

Since this event lasted longer than one night, technicians had time to disconnect the batteries powering the telegraph systems to avoid the risk of any further damage. Surprisingly, they discovered they could still send telegraph messages without any connected battery power due to the lines still remaining electrically energized! Ship's logs recorded how the skies above the oceans turned dark red throughout the night while waves and storms lashed the ships. Ship's compasses would constantly spin making navigation impossible. [1]

At the same time this solar storm was playing havoc on the telegraph systems, an English astronomer recorded in his field notes that he had observed a very bright pulse of light leaving the center of one of the black areas on the sun's surface the day before. Since this solar storm was moving through space much slower than the speed of light, it took eighteen hours for this energy field to reach the earth's path after he first observed the bright light leaving the sun's surface.

Richard Christopher Carrington was the English astronomer and Secretary of the Royal Astronomical Society at the time, using an observatory he built himself near Surrey England. He was the first person to realize the past two nights of glowing skies and the damage to all telegraph systems must be connected to the major disturbance he had observed on the sun the day before. This is known today as the Carrington Event of 1859 in honor of his discovery.

This 1859 event provided a real-world example of what a CME or solar storm can do to any electrical equipment connected to long power and phone lines, and why all of the labor-saving electrical devices and electronic communication systems we now take for granted, and our everyday lives depend on, will fail if this level of electromagnetic energy struck the earth today.

On March 9, 1989 a more recent CME was observed leaving the sun's surface. Unfortunately, while less concentrated than the 1859 Carrington event, the more northeastern New England states and Quebec Canada received a glancing blow when this burst of electromagnetic energy struck this part of the earth that early morning.[2] This caused significant damage to high-voltage transmission lines serving Quebec's hydro-electric system, plus tripped safety devices offline and destroyed several high-voltage transformers in both Canada and the grid interconnected state of New Jersey.

All of Quebec plus six New England states went dark immediately. Other electric utilities located in the United States that served these northeastern states also experienced damage and power outages although located further away from the main impact point. The damage was estimated at over six-billion dollars. [3]

More recently, on July 23, 2012 a Coronal Mass Ejection (CME) estimated to be greater than the disastrous 1859 Carrington event, crossed the earth's path only nine days before the earth passed through that same exact point in space. No doubt if this crossed the earth's path nine days later, it would have virtually destroyed most, if not all of the electric grid, communication systems, and electrical equipment over the entire half of the earth facing this energy stream when it crossed. Not surprisingly, the public was not made aware of this extremely close call until it was reported by NASA over two-years later!

While most solar events are not as strong as the 1859 Carrington Event, the earth still passes through much smaller "solar storms" ejected from the sun on a fairly regular basis. Since even these small solar discharges can still disrupt electrical and communication systems for days here on earth, we now have advanced warning observation satellites specifically trained on the sun, and the National Oceanic and Atmospheric Administration (NOAA) now operates the official space weather prediction center. It can take from several minutes to a day or more from the time an observable CEM leaves the sun to reach the earth. This delay provides time to send out an early warning e-mail to utility and satellite communication operators, along with a severity rating that helps predict if the event will do minor or major damage.

Like hurricanes and tornadoes, these geomagnetic solar storms have their own classification system from G1 to G5, with G1 usually indicating a solar storm is approaching that will probably cause minimum problems. At the G5 level of solar activity a CME level event is anticipated, scientists predict the stabilizing systems in earth orbiting satellites will be damaged causing them to drop out of orbit and fall to earth; passenger planes could crash when their navigation and "fly-by-wire" computerized controls fail; and our electrical grid and communication systems will suffer catastrophic damage.

While a coronal mass ejection is not the same as an electromagnetic pulse from a nuclear explosion, the damage to our electrical grid will be similar, which is why I have included this discussion in a book primarily discussing EMP. Actual testing has shown how protective devices designed to block the high voltage spikes from a nuclear EMP event will also block the high-voltage spikes in the electric grid induced by a solar CME, but the reverse is not true. The energy wave of a solar CME is much slower to reach peak voltage than it takes for an EMP event, but a CME does cause the same damage to high-voltage transformers as the E3 phase of an EMP event.

As stated previously, a solar CME passing directly across the earth's path at the same time the earth intersects this point is rare. However, scientists predict the odds are turning against us and we may be long overdue for the next "big one," and it will be devastating. Statistically there is a 12 percent chance of a CME directly impacting the earth during any period of maximum sunspot activity, which occurs every eleven years. Some electric grid operators are reluctantly considering adding additional protection to high voltage transformers to lessen the damage from a solar storm or CME. Unfortunately, these modifications are not sufficient to protect against an EMP. In chapter 6 I will go into more detail concerning the battle between government agencies and the grid operators, who apparently would prefer to do nothing at all.

CHAPTER 5

EMP Impact on the Power Grid

The primary concern for an EMP attack is how it can take down our nation's electric grid which can last months, if not years. However, before we get too deep into the science of how an EMP and solar storm event can damage the United States electric grid, we need to review just how vulnerable our electric grid is today.

The United States electric grid consists of over three thousand separate investor-owned utilities and rural cooperatives, operating over nine thousand generating facilities, and *none* are under the direct control of any government agency. These are totally independent for-profit utility corporations and co-ops with their own network of generating plants, substations, and transmission lines. While many are now interconnected to each other to provide redundancy and additional peak demand capacity for their customers, they are still separate corporations with separate management, separate control centers, and separate maintenance groups. Unfortunately, history has shown a major disruption to one utility grid operator can cause a cascade-type failure event to many others.

The national electric grid is divided into three basic sections, with the Western Interconnection and the Eastern Interconnection sections separated between Montana and North Dakota. This dividing line extends down to Texas, which operates under its own grid system that is separate from all other grid systems. While individual grid operators in any given section of the country are interconnected to each other, there is limited interconnection between these three major divisions.[1] If half of the country sustains a total grid down event, it will be physically and electrically impossible for the remaining half of the country to serve as a backup power source for the failed system until it is back in operation.

Over eighteen hundred independent grid operators are members in the North American Electric Reliability Corporation (NERC) which establishes their own interconnect standards and security protocols for its members. Notice the word "corporation" in the name, as this is a totally voluntary trade organization with no enforcement powers on its members. At the federal level, the Federal Energy Regulatory Commission (FERC) can establish grid reliability standards, but this government agency is still required to work through the voluntary NERC corporation and it, not the Federal (FERC) agency, decides how best and when to implement any federal mandated changes for the grid operators. In 2016, the voluntary NERC Corporation tried to implement twenty-five increased security measures on all member utilities, but to date none have been implemented.[2] The relationship and ongoing battles between FERC and NERC are legendary, and I will address it in much more detail in chapter 6.

Fig. 5-1. High-voltage power lines carry power from distant power plants to regional substations.

Each of the independent grid operators manage their own system of high-voltage transmission lines and substations crisscrossing our country, and much of this equipment was installed in the 1960s and is now over sixty years old. Since 2008 there have been 552 fully operational coal-fired power plants spread across the country that have been scrapped due to their inability to meet new EPA emission standards, and the recent lower cost for national gas verses the price of coal. To date only one new replacement plant has been built.[3] There are a total of fifty-eight nuclear power plants operating in the United States, with an average age of forty years as of this 2020 publication. The oldest nuclear plant still in operation is Nine Mile Point in New York completed in 1969. No doubt as these nuclear plants and remaining coal-fired plants continue to age, there will be additional loss of system capacity, and these plants are not being replaced. This aging electric infrastructure plus reduced generating capacity has made the United States electric grid much more susceptible to electrical stress and overload than any time in our past.

The reason given why most of the independent electric utilities have not taken any corrective action to prevent EMP damage is cost. The United States government can establish EMP guidelines and make recommendations but has no real enforcement capability to determine what these independent utilities must do. The thousands of separate grid operators insist the security upgrades to reduce the risk from an EMP attack or solar storm will cost billions of dollars. In addition, the electric utility industry likes to point out these expensive upgrades are primarily to protect the nation against a potential military attack against the United States.

Operators feel any system upgrade costs to reduce the impact of a nuclear attack should be paid by the government and taxpayers, not their own individual customers. No doubt this debate will continue right up to the day of an EMP attack, then the finger pointing will begin while half the country will be without electricity for a year or more.

EMP IMPACT ON THE POWER GRID

Fig. 5-2. One of over 55,000 high-voltage substations scattered across the country.

The continental United States has 360,000 miles of high-voltage transmission lines operating at over 115,000 volts, including over 71,502 miles of extra high-voltage lines over 345,000 volts. These lines supply 2,146 extra high-voltage transformers (345 kV or higher). These high-voltage networks supply over 2 million additional miles of lower-voltage distribution lines connecting 55,000 substations that supply individual cities and industrial plants.[4]

The average age of all high-voltage transformers in the United States is almost fifty years, with 70 percent over twenty-five years old. Many were installed between 1950 and 1970 when the grid was greatly expanded after WWII, but there are still some high-voltage transformers in operation that are over seventy years old.[5] The average expected design life of a high-voltage transformer is forty years, which indicates the grid operators have continued to delay what are expected to be massive replacement costs as these systems starts to fail in mass, even without an EMP or solar storm!

The vulnerability of the electric grid system from an EMP or solar storm will be the damage they can do to the high-voltage transformers scattered across the country. All transformers, regardless of size, are alternating current (AC) devices, which can overheat and quickly fail if subjected to direct current (DC) power flows superimposed on the long power lines and into these transformers by an EMP or solar storm event.

As covered in chapter 4, a solar storm can impact the earth with the same damaging electromagnetic energy as the E3 late time phase of an EMP. Utility meters indicated the geomagnetic storm that struck eastern Canada on March 13, 1989, generated over 100,000 amps of ground-induced electrical current down the power lines and into multiple high-voltage transformers. This drove these high-voltage transformers into "saturation" which created harmonics that caused the transformers to overheat and actually melt.[6]

This 1989 solar storm caused the partial melt-down of all three single phase transformers rated at 406 Megavolt Amp (MVA) which were located in the Salem, New Jersey nuclear power plant. These three step-up transformers supply the 500 KV distribution transmission line leaving the plant's number-one reactor section. Inspectors found severely damaged internal wire insulation, and multiple masses of solid copper that solidified below the melted copper coils above. Remember, this was caused by the sun!

Surprisingly, on September 19, 1989, just six months later, a second solar storm struck and caused almost the same damage to all three step-up transformers in the unit no. 2 reactor section in the same Salem, New Jersey facility.[7] The Salem nuclear plant is located on the banks of Delaware Bay just off the Atlantic Ocean. Due to being built over former marsh land, it is supported by hundreds of one-hundred-foot-long steel pilings. Solar storms historically cause the most damage near edges of large bodies of water, and these pilings made an excellent path for the induced currents to be grounded out by the earth.

EMP IMPACT ON THE POWER GRID

Follow-up studies found similar sized transformers in eleven other nuclear power plants that are part of the same grid system also suffered eventual failures over the next two-years. Since the initial damage was less severe due to the greater distance, it took time for the breakdown of the internal wire insulation to finally create hot spots which eventually shorted out the coil windings.

As I noted in chapter 4, a solar storm produces almost the same type geomagnetic disturbance to long electric lines as the E3 phase of an EMP but does not have the earlier E1 and E2 phases. Even this relatively small solar storm that mainly struck the Quebec area of Canada still totally destroyed multiple high-voltage transformers as far away as New Jersey. This proves we can expect far more damage to the thousands of high-voltage transformers making up the United States electric grid if we suffer a more direct hit.

I noted in chapter 4 solar storms historically occur when there is major sunspot activity, which is typically every eleven-years. While there are times when the spacings between these peak periods have been several years or more, this is still a fairly regular cycle. This would mean we are long overdue! A real EMP event would not only surpass the grid damage of a solar storm but would also include the major damage to all vehicles, and all electronic and communication systems due to both E1 and E2 energy, which are not generated by a solar storm.

High-voltage transformers in the United States are aging, and this makes them even more vulnerable. For example, transformers are basically large coils of copper wire wound around many layers of very thin steel plates, specifically manufactured to have strong magnetic properties, and held tightly together by extremely large bolts. As transformers age, the insulation on the copper coils can start to breakdown, core assembly bolts can stretch, and these thin metal plates will start to separate. Under normal conditions, this will cause additional heating of the transformer core and mechanical vibration, even without the damaging effects from a EMP or solar storm.

In 2013, a simulated solar storm test conducted by Idaho National Laboratory was performed on a 138,000-volt AC transformer that was sixteen feet wide, twenty-five feet long, and twenty-five feet high. This 150,000-pound transformer started vibrating so hard during this test that it actually started to bounce across the concrete floor of the lab! It should be noted the level of simulated DC voltage they were injecting into this AC transformer was far below what a real EMP or solar storm would produce. This test had to be stopped when the large transformer started to actually move for fear of damaging the building structure and the multimillion-dollar transformer.[8]

There are currently less than a handful of manufacturers of extra high-voltage transformers having a 345 kV or higher capacity rating manufactured in the United States, and all are actually owned and operated by foreign companies. These extra high-voltage transformers take up to two years to design, build, and ship. Transformers in this size range are as big as a house and weigh up to 820,000 pounds. Customized trailers just to move a transformer this size are over seventy feet long, require more than two hundred wheels and axles, and take up two lanes of the highway.[9] Just to move from a coast port to a final installation location usually requires reinforcing bridges and raising and disconnecting electric and phone lines crossing their intended route.

EMP IMPACT ON THE POWER GRID

Fig. 5-3. Extra high-voltage transformers can weigh over 820,000 pounds and are as big as a house.

These transformers are designed for a specific substation and each will have different voltages and different power capacities. Their large insulated terminals will be different sizes and have different mounting orientations. Costing in excess of $8 million with each having a unique design means there is no standard stock transformer design and no spare transformers this size just sitting on somebody's loading dock waiting to load up and head to a failed substation!

Transformers are not the only damage an EMP or solar storm can do to the electric grid. In addition to these large transformers, substations require large high-voltage capacitors, transfer switches, insulators, and relays. These are also custom switchgear components having long lead times and costing hundreds of thousands of dollars each. These can suffer cracked insulators, welded or failed switch contacts, and destroyed switch controls if subjected to the extremely high voltages and currents produced by an EMP or solar storm.

China has an electric grid almost the same size as the electric grid in the United States yet manufactures all of their own high-voltage transformers and switchgear and do not import transformers from any other countries. For comparison, the United States only manufacturers 15 percent of our yearly transformer needs, and these are primarily intended for system expansion, not replacement of aging units. The combined production output of all manufacturers in the entire world is just over one hundred extra high-voltage transformers each year to supply the needs of all other countries.

This means the United States is going to be standing in a very long and very slow moving line of world customers to receive replacement transformers if we suffer the simultaneous loss of thousands of these transformers. This fact coupled with the average two-year design and manufacturing cycle time, plus the nightmare delivery and installation issues, will make it very difficult to bring the United States electric grid back to normal after an EMP or solar storm. Out of the fifty-five thousand high-voltage substations, if only nine carefully selected substations are destroyed at the same time, it's conceivable up to half of this country could be without electric power for a year or more regardless if it's caused by an EMP, solar storm, or sabotage.[10]

Based on experience from numerous past grid blackouts caused by storms or equipment failures, it requires real-time constant communications between the centralized grid operators and the remote power plant operators located in many different parts of the country to coordinate and gradually bring a failed grid system back to normal operation. If phone and internet communication systems are also damaged by an EMP event, it will be almost impossible for operators to safely restart a highly interconnected power grid, especially with the demolition of so many coal-fired powerplants that are no longer available to provide this cold start capability. This also does not consider the communications required to coordinate the ordering, building, shipping, and installation of each replacement transformer.

EMP IMPACT ON THE POWER GRID

Originally each separate electric utility owned and maintained their own power plants, transmission lines, transformers, power distribution systems, customer meters, and monthly billing operations. However, after many states deregulating their electric industry back in the late 1990s, the generating plants and the transmission lines are now owned and maintained by totally separate corporations. What happens after an EMP attack damages many of these separately owned parts of a grid system?

Let's say the entire electric grid serving the eastern half of the country has just been taken down by an EMP, substation sabotage, hacking of the computer centers, or some combination of all three. What if the independent operators of the cross-country high-voltage transmission lines connecting all these separate power plants and substations do not agree with the manual procedures being planned by the plant operators to re-energize their individual distribution lines? Who is responsible if these cross-country lines are damaged or multiple large transformers are destroyed due to the attempt to operate the system manually when they were designed to be totally controlled by a very sophisticated computer system that has now failed? It's not uncommon for an overloaded power line to heat up, stretch to the ground, and catch fire.

And who is going to make all these repairs to these substations, transformers, and power plants scattered across the country? These thousands of utility workers and their home life will suffer the same hardships as the rest of us. Will their service vehicles even start, and will they be able to find fuel during the many months it will take to make these repairs? Where will they eat and sleep for weeks at a time away from home if all restaurants and hotels are also without power and closed? What about the manufactures of all those electrical repair parts? What about shipping and delivery drivers? All this will be occurring while most areas of the country will be without electric power, limited food deliveries, fuel shortages, and no credit cards, e-mail, or cell phones.

Other than a government war powers act or a military takeover of these independent power plants, substations, and transmission lines, there is no one central control group above these independent utility operators. The financial costs and liability involved in getting these separate systems operational and interconnected again could be astronomical. Without the normal computerized safety procedures and controls, system repairs and a cold restart could easily take months and could even require military and congressional action considering the major liability risks involved. Yes, even after a catastrophic EMP attack or solar storm event, we all might be waiting a year or more by candlelight and eating beans and rice, while utility lawyers argue endlessly as to who is going to pay! Remember, these are privately owned for-profit businesses and their management answers to their stockholders, not the government.

CHAPTER 6

EMP Inaction by Government and Grid Operators

Considering the consequences of an EMP attack have been known since the nuclear testing in the 1950s, you would think by now our government would be demanding the hardening of the electrical grid. You would be wrong! Before getting into the main points of this chapter, I want to present to you a list of the numerous EMP studies that have been completed since 1997 by multiple government agencies yet, to date, almost nothing has been done.

Note the titles of each report, many of which are hundreds of pages long and summarize years of analysis by some of the best nuclear scientists and engineers in this country. All of these reports—and this is only a partial list—include specific recommendations to secure the electric grid from both a military EMP attack and a solar storm. So why hasn't anything been done if we already know what the problems are?

These detailed EMP studies include:

1997 "Threat Posed by Electromagnetic Pulse (EMP) to U.S. Military Systems and Civil Infrastructure" by US House of Representatives

1998 "High Altitude Electromagnetic Pulse (HEMP) Protection for Ground Bases Facilities Performing Critical, Time-Verses Missions" by Department of Defense

EMP INACTION BY GOVT AND GRID OPERATORS

2004 "Report of the Commission to Assess the Threat to the United States From Electromagnetic Pulse (EMP) – Volume One" by U.S. Congressional Commission

2008 "Report of the Commission to Assess the Threat to the United States from Electromagnetic Pulse (EMP) Attack" by U.S. Congressional Commission

2012 "Large Power Transformers and the U. S. Electric Grid" by U.S. Department of Energy

2013 "Electric Grid Vulnerability" by U.S. House of Representatives

2014 "Physical Security of the United States Power Grid: High-Voltage Transformer Sub-Stations" by Congressional Research Service

2014 "Securing the U.S. Electric Grid" by Center for Study of the Presidency and Congress

2016 "Electromagnetic Pulse (EMP) Protection and Restoration Guidelines for Equipment and Facilities" by United States Department of Homeland Security

2017 "Report to the Commission to assess the Threat to the United States from Electromagnetic Pulse (EMP) Attack" by Dr. William R. Graham, Chairman

2017 "Nuclear EMP Attack Scenarios and Combined-Arms Cyber Warfare" by Dr. Peter Pry

2017 "Call to Action for America" by Task Force on National and Homeland Security, Secure the Grid, and Other Partners

2018 "Electromagnetic Defense Task Force Report" by Curtis Lemay Center for Doctrine, Maxwell Air Force Base

2019 "Surviving a Catastrophic Power Outage" by the President's National Infrastructure Advisory Council

They each highlighted the United States electric grid's extreme vulnerability to an EMP event, and discuss what will happen when, not if, a devastating EMP event occurs. These risks are well known and have been extensively documented by over thirty-years of studies. However, with the exception of a few states forcing the utilities in their jurisdiction to make upgrades, as of this writing none of these recommendations have been implemented by the balance of all electric grid operators. In addition, other than suggesting more studies, no real corrective action has been taken by the multiple government agencies established to monitor the grid and protect us from an EMP attack.

Unlike the lack of support for the EMP Commission's recommendations by previous administrations, President Trump did recognize the damage an EMP can cause to our way of life. In March 2019, he issued an executive order titled "Coordinating National Resilience to Electromagnetic Pulse." This order requires specific government agencies in the administration to submit documentation regarding the anticipated effect to their agency of an EMP attack.

The Department of Homeland Security was given twelve months to identify how critical infrastructure could be strengthened to withstand an EMP attack. Unfortunately, both the Department of Homeland Security and the Department of Energy continued to ignore the urgency of implementing the recommendations of the EMP Commission. Things may be starting to change, however, as President Trump signed into law in December 2019 the National Defense Authorization Act, which includes as law the president's March 2019 executive order on EMP. Time will tell.

While the slow-moving federal government finally starts to get its EMP protection plan together, you may be surprised to learn at the state level there are a number of governors and state legislators that do take this EMP threat very seriously and have stopped waiting for any direction from the federal level. As of 2020, the states of Arizona, Florida, Oklahoma, Maine, and Virginia have introduced legislation or have initiatives in place to incorporate hardening the electric grid in their respective states against the risk of an EMP and solar storm.[1]

If the electric utilities continue to fight against any EMP protective legislation coming out of Washington, DC, they may soon face a hodgepodge of different regulations and competing standards being dictated at the state level. It's clear EMP is a real concern for many citizens, even if the bureaucrats and elected representatives in Washington continue think the problem will just magically go away.

In 1997, a congressional sub-committee established the Commission to Assess the Threat to the United States from Electromagnetic Pulse (EMP) Attack. In 2004, the results of this sub-committee's hearing were a detailed five-volume report that summarized the vulnerability of the United States to an EMP attack. While much of this report is still classified, the summary findings stated, "our increasing dependence on advanced electronic systems results in the potential for an increased EMP vulnerability of our technologically advanced military forces which makes use of an EMP weapon by an adversary an attractive option".[2] Translated into laymen's terms, almost everything that makes our lifestyle possible requires microelectronics to work, and these systems are extremely vulnerable to an EMP attack.

In 2008, this subcommittee released its final report titled "Report of the Commission to Assess the Threat to the United States from Electromagnetic Pulse (EMP) and briefed the United States House Armed Services Committee with a detailed summary of their findings. Here is a brief excerpt of their findings:

> Should significant parts of the electrical power infrastructure be lost for any substantial period of time, this commission believes that the consequences are likely to be catastrophic, and many people may ultimately die for lack of the basic elements necessary to sustain life in dense urban and suburban communities. This commission is deeply concerned that these impacts are likely in the event of an EMP attack unless practical steps are taken to provide protection for critical elements of the electric system and for rapid restoration of electric power, particularly to essential services.[3]

If I may translate again, if you live in a densely populated area, you could likely die when, not if, civilization breaks down into anarchy after an EMP attack.

In June 2010, the House introduced H.R. 5026 titled the Grid Reliability and Infrastructure Defense Act, or GRID for short. The purpose of this bill was to allow the Federal Energy Regulatory Commission (FERC) to develop new utility industry standards that would protect critical areas of the utility infrastructure from both cyber and EMP attack. Although this bill passed the House, it was never allowed out of committee in the Senate, where it died.

In February 2011, the House introduced HR. 668 which was the Secure High-Voltage Infrastructure for Electricity from Lethal

Damage (SHIELD). The purpose of this act was to direct the Electric Reliability Organization to develop standards for large power system operations that would increase their ability to withstand solar geomagnetic storms and EMP attacks. This act has been referred back to a committee. Are we seeing a pattern here?

What if a low-yield nuclear bomb was detonated several hundred miles above the center of this country and produced massive damage to the electric grid? What if this attack was carried out by a rogue nation or terrorist group and not a major military attack by a known country?

If an EMP attack on the United States did not kill anyone or destroy a single building, would we actually respond with a nuclear counterattack? And if not nuclear, are we willing or even have the ability to respond with a massive non-nuclear response? It has been estimated the long-term loss of our electric grid would eventually cause the deaths of over 90 percent of the United States population though starvation and medical complications.[4]

The Department of Homeland Security has studied every possible type of threat that could possibly happen to this country and has developed a master response plan that includes a total of fifteen high-risk possibilities. These include everything from terrorist acts and natural disasters to domestic bombings.

Even though the congressional EMP committee tried to convince the Department of Homeland Security to add the possibility of an EMP attack and solar storm to this response plan, the final report from Homeland Security did *not* include the risk of an EMP or solar storm in this master plan! The Homeland Security Administration has yet to consider EMP as a potential threat to this country and has not completed any planning scenarios for anything associated with an EMP or solar storm threat.[5] I have been told the main reason these EMP disaster scenario exercises were stopped was because the team members could never find a viable solution!

EMP INACTION BY GOVT AND GRID OPERATORS

It must be understood that in order for a high-altitude EMP (HEMP) to cause infrastructure destruction to a wide area, the nuclear bomb must be detonated several hundred miles above the center of the United States. This requires the use of a missile capable of carrying the nuclear payload to a specific location from a very great distance. However, if fired from a ship in the Gulf of Mexico or along our eastern seaboard, our early warning systems will have only seconds to identify and track. Once located, the only way to disable either a missile or nuclear EMP weapon is to blow it up prior to detonation, which is impossible if the missile only takes seconds to basically go straight up.

We currently have no way to stop an EMP attack, and our government has not imposed any new regulations or expressed any public concern that would cause our electric and communication industries to wake up to the threat. Regarding our military readiness, the Department of Defense concluded that all critical missions, are entirely dependent on the national electric grid, and this infrastructure is commercially owned, not government or military owned. In addition, 99 percent of all electrical power consumed by all military bases and airfields originates outside the fence.[6]

Every military base I have visited, and that includes many across the United States and in Europe, have large electrical substations located just outside their perimeter fences since all electric power for the base is connected to civilian high-voltage transmission lines supplied by distant commercially owned power plants.

Each military base is like a self-contained city and typically includes an airfield, base housing, training facilities, multiple cafeterias, medical facilities, transportation systems, fire stations, maintenance facilities, and even recreational facilities, which support thousands of both active duty and support personnel. All electric power, however, is supplied by the same national electric grid that supplies the rest of us. While a large number of individual base facilities do have their own emergency backup generators, most are building specific and consist of smaller diesel generators with limited emergency fuel supplies. Even a temporary grid outage can cause a serious reduction to the operation and readiness of these military bases and airfields.

A takedown of our electric grid by an EMP or solar storm would be catastrophic to our military's ability to respond or function. What good is hardening the military equipment attached to one end of a power line or phone line inside a military base or airfield if everything supplying power and communications to these lines are civilian owned using non-hardened commercially manufactured hardware?

From a brief review of all congressional studies, hearings, and over thirty years of research into EMP, the question remains: Why haven't the grid operators implemented any of these protective recommendations? The simplest answer is, there is no government agency with the power and authority to dictate what the thousands of independent grid operators must do to protect their systems from an EMP attack or solar storm!

The primary government agency that deals with national utility issues is the United States Federal Energy Regulatory Commission (FERC). Unfortunately, this federal agency is not allowed by its founding legislation to deal directly with these thousands of individual electric utilities. Instead, any proposed changes to the regulations regarding the operation and security of the electric grid must first go through the North American Electric Reliability Corporation (NERC).

This name sounds impressive but note the word *corporation* in the title. The NERC is a non-governmental trade association made up of all grid operators. History has shown they do not like to change anything in their day-to-day operations, especially if it will increase their operating costs. While the Federal Energy Regulatory

EMP INACTION BY GOVT AND GRID OPERATORS

Commission (FERC) can request the board of the NERC to consider their proposed recommendations to better protect the grid from an EMP, the NERC board can modify, water down, or outright reject any of these government requests. Basically, the electric utility industry is the only critical industry in this country that is self-regulated.[7]

In contrast, the Federal Aviation Administration can dictate safety features and the design of all aircraft and airport runway systems; the Federal DOT can dictate safety features engineers must incorporate in all highway and vehicle designs; The Federal Environmental Protection Agency dictates what every industry can discharge into the air and water down to what a farmer can use for irrigation, The Federal Nuclear Regulatory Commission licenses and dictates the design requirements for every nuclear reactor, and the Federal Food and Drug Administration regulates the manufacture and labeling of our foods and drugs, and can even impose criminal penalties for the distribution of unsafe products.[8] Therefore, why is the federal agency responsible for ensuring the entire electric grid is safe from an EMP or solar storm only allowed to "suggest" what the electric utility industry should consider regarding increased reliability?

Who decides how much protection high voltage transformers should include in their design and what types of neutral grounding devices should be added to block the damaging electromagnetic energy trying to enter into each transformer from an EMP or solar storm? These devices have proven to be an effective way to block large geomagnetic ground currents created from a solar storm, which are very similar to the effects from the E3 phase of an EMP event.

Currently the NERC representing all grid operators have relied on computer models to decide which high-voltage transformers in their networks will require adding grounding protection, and what voltage threshold will be needed. You can't just destroy multiple

EMP INACTION BY GOVT AND GRID OPERATORS

transformers the size of a house and costing millions of dollars while testing to find the right voltage and current limits! It's also not easy to recreate the ground-induced currents into a test grid expected to be generated by an EMP or solar storm unless you happen to have an extra nuclear bomb laying around. So, for now, all transformer protection equipment is based on computer modeling.

However, as noted in chapter 4, there was an actual solar storm event in March 1989 that hit northeast Canada and did extensive damage to Canada's electric grid including several northeastern states in the United States that are interconnected with Canada's grid. Multiple high-voltage transformers were damaged or destroyed, and since everything was metered, there was reliable data available to "fine-tune" these computer models. Unfortunately, NERC decided not to use any of this real-world data, even though these transformers experienced voltage surges and ground currents four to seven times higher than what the computer programs estimated or should be used in transformer protection design requirements!

No doubt, to minimize costs, NERC and all grid operator members want to use their existing lower computer-generated damage estimates, which to date indicate *all* of their existing high-voltage transformers have adequate protection and do not need to add any costly grounding and neutral bonding devices! The electric utility industry has at least started to consider the effects of a solar storm, which they consider to be far more probable than an EMP. When designing protective devices to protect a transformer from a solar storm event, their computer estimated voltage limits are multiple times *lower* than required to protect the same transformer from an EMP.

If these safety devices were sized to protect against the higher levels of voltages and currents generated by the E3 portion of an EMP, it would be more than adequate to protect against damage from even the highest recorded solar storm. While the NERC and

EMP INACTION BY GOVT AND GRID OPERATORS

its members continue to drag their feet for doing this minimum level of protection against a solar storm, they are not even talking about the potential risk from and EMP. [9]

Other signs the grid operators are not willing to upgrade anything is the electric industry's NERC published in 2012 several reports that significantly underestimated the effects to the grid from a nuclear EMP and solar storm. This report generated very inadequate standards for protecting the grid, and undermined multiple state EMP initiatives already in the planning stage which specified much stronger protection.

In 2016 the electric power industry funded the Electric Power Research Institute (EPRI) to publish a companion report that also significantly underestimates the potential damage to the electric grid by a nuclear generated EMP. In addition to grid operators, many other industrial groups who would also need to do costly upgrades, are using these same reports to help them lobby against all state and federal efforts to better protect the grid. [10]

To be fair, most grid operators feel an EMP event will involve a military attack against the United States, so any defense preparations needed to protect the United States should be at the expense of the government and taxpayers, and not their rate payers. Regardless, this debate is ongoing and deciding who ends up paying for all this is above my pay grade.

CHAPTER 7

EMP Impact on Our Military

During early testing, nuclear scientists discovered that when an electromagnetic shockwave is produced by a nuclear bomb blast, it would cover a very wide area if the detonation occurred at a high altitude. This is called High Altitude EMP (HEMP) and causes the most damage when the detonation is over two hundred miles above the earth's surface. At this height the radiation and explosive forces will not reach the earth's surface, but there will be massive gamma and X-rays entering the upper atmosphere producing the very large EMP wave as discussed in chapter 3. If a nuclear device was detonated above the central United States at this elevation, the effects of the EMP will reach the entire continental United States from coast to coast.

A HEMP weapon is somewhat challenging to develop, but it's possible that both a nuclear warhead and a missile can be separately acquired by a rogue nation or even stolen, then joined together. It is estimated over 128,000 nuclear warheads have been assembled since the beginning in 1945.

During the cold war the United States lost and are still missing eleven nuclear warheads, including an eight thousand-pound nuclear bomb lost in 1958 near the coast of Georgia, which was jettison from a B-47 bomber; a twenty-four-megaton nuclear bomb lost off the coast of North Carolina in 1961 from a crashed B-52 bomber; a 1-megaton hydrogen bomb lost off the coast of Japan in 1965 from a sunk E-4E bomber; a nuclear warhead of unknown size off the coast of Greenland in 1968 when a B-52 bomber crashed; not to mention

EMP IMPACT ON OUR MILITARY

the nuclear attack submarine, *Scorpion*, which sank in 1968 near the Azores Islands with two nuclear bomb-tipped missiles.[1] While *Scorpion*'s two nuclear missiles are technically not lost, as of this day they have still not been recovered. This is just the nuclear weapons lost and still missing by the American military. It is estimated another 100 or more nuclear weapons were lost during the cold war by other countries, not to mention the 250 nuclear suitcase bombs manufactured by Russia during this same time period and over 100 are still unaccounted.

Perhaps a terrorist getting their hands on a few pounds of plutonian or enriched uranium may not be as difficult as we have been led to believe? Radioactive materials from lost weapons will still be perfectly suitable to make new nuclear weapons if recovered, but the associated non-nuclear explosives, detonators, and fusing triggers that actually cause the nuclear chain reaction will have deteriorated and these components are now useless.

While many of these non-nuclear components are available on the black market, actually constructing a nuclear bomb still requires some very sophisticated machining and technology not possessed by the average terrorist. However, strapping these nuclear materials to conventional explosives could render a city uninhabitable for the next one-hundred years due to radiation from the scattered debris. While this will not cause an EMP, the damage will still be unacceptable.

Thirty different countries have an estimated ten thousand missiles capable of lifting a nuclear warhead above the United States.[2] A South Korean defense ministry report estimated North Korea had manufactured and shipped over four-hundred relatively inexpensive SCUD missiles to multiple middle eastern countries including Syria, Iran, Libya, and Egypt, and they have another five hundred in inventory, plus over one thousand SCUD missiles deployed.

EMP IMPACT ON OUR MILITARY

The newer SCUD-B and SCUD-C missiles are capable of speeds in excess of Mach 5 and can reach altitudes over 115-miles up. Since their total travel range is less than five hundred miles, they would need to be launched from a ship near our shores and aimed almost straight up to reach anywhere near the upper ionosphere. While not very accurate when trying to hit a specific target on the ground, accuracy is less important when the goal is just to lift a nuclear warhead as high as possible. In relation to thousands of SCUD missiles held by multiple countries, the relatively poor security at multiple storage facilities for these missiles makes these ideal candidates to be stolen or sold on the black market.

In other words, it's not unreasonable for a rouge agency or terrorist group to obtain and combine a separate nuclear warhead and missile, with or without the support of their host country. This would be an ideal weapon when the goal is not to destroy a country's infrastructure, which could be used for an occupation that was sure to follow. There would be no ground-level mushroom cloud, and no slow-killing radioactive fallout. All people, buildings, facilities, highways, bridges, and animals would still be here, except many electronic devices would no long work and there would be widespread power outages and a communication blackout. Many cars, trucks, trains, planes, helicopters, and construction equipment would experience some level of damage to their electrical or engine ignition systems. Some vehicles would stop dead, never to run again, and some may not be affected at all. Chapter 11 deals specifically with the effects of an EMP on vehicles.

Contact with friends and family in another city or state would be impossible, at least initially. There would be very limited television broadcasts as many stations will be off air due to EMP damage, and only a few radio stations having emergency backup power will still be broadcasting if not damaged. For the first few days the over-energized ionosphere would severely limit radio wave propagation.

EMP IMPACT ON OUR MILITARY

Calling the police or fire departments, ambulances, and other 911 emergency services, assuming a phone still works, would be useless. Emergency communication systems would also be either temporarily disabled or destroyed. You would be on your own, so you would need to start walking to get out of a congested area. In a city a bicycle or your shoes may be the only transportation still operational and able to make it through traffic gridlock that will occur immediately after an EMP attack.

Whenever a power outage affects a large area of the country, regardless of the cause, grid operators controlling the disabled power plants still require external power from backup generators or unaffected parts of the country to place these disabled systems back in operation. This is called a black start, and this allows powering all of the needed plant controls, voltage stabilizing systems, and computer communication networks while getting ready to begin a careful and gradual reintroduction of power back into the grid.

In addition to needing a working grid in an adjoining and undamaged area of the country to power a local system black start, all nuclear power plants require a continuous supply of power twenty-four hours per day from external sources throughout the entire time a nuclear plant is shutdown. This is needed to operate their large pumping systems that must continually circulate cooling water through the nuclear cores to prevent a meltdown. While all nuclear plants are designed with backup generators to keep these cooling systems and their computerized controls operating during a plant shutdown, these backup systems do not have an unlimited supply of fuel. Since an EMP or solar storm event could disrupt all normal fuel deliveries due to traffic gridlock, loss of traffic controls, and closed gas stations, it is extremely doubtful these backup generators would be able to operate indefinitely.

Although not directly related to an EMP event, perhaps this is a good time to review the Fukushima Daiichi nuclear power plant disaster that occurred in March 2011, when an earthquake followed by an ocean wave tsunami struck the island of Japan. This plant's six nuclear reactors with a combined capacity of 4.7 gigawatts is considered one of the world's largest nuclear power plants and is located right on the coastline of Japan. Redundant safety control systems did immediately shutdown three of the six nuclear reactors that were operating at the time, when seismic sensors were triggered by the initial earthquake.[3]

The magnitude of the earthquake and the resulting tsunami wave that struck the power plant both exceeded the plant's original design limits that were in effect forty years ago when the facility was constructed. This inundation sent sea water surging into the six reactor areas, plus there was physical damage from the initial earthquake to several parts of the containment vessel. This caused a loss of reactor controls and power for the multiple cooling systems. Cooling water must be pumped through each reactor core at all times, which can quickly reach several thousand degrees if this water flow is stopped. This high temperature will easily melt metals and burn through the bottom of the thick metal and reinforced concrete containment vessel, which will release the contaminants down into the water table in the earth below.

In addition, a separate open pool of water located outside the reactor, but still inside the large containment vessel, is used to store hundreds of spent fuel rods that have been accumulating during multiple years of refueling the reactors. If the water level in this pool is not constantly circulated or the water level drops, the exposed metal jackets enclosing these fuel rods can overheat and split open, causing the formation of explosive hydrogen gas. Obviously, it is extremely important for any nuclear plant to keep these multiple cooling and pumping systems fully operational twenty-four hours per day during the entire time the plant is shutdown, regardless if this shutdown was for normal repairs or major damage. If the grid is down, backup power to operate these cooling systems is provided by multiple backup diesel generators, each having excess capacity in case not all generators are operational.

Unfortunately at Fukushima all of these backup generators were located at ground level outside the sealed containment vessel and were severely damaged by the earthquake and resulting tsunami wave, leaving both the nuclear reactor cores and the fuel rod storage pools without any means to keep safely cooled. The resulting multiple hydrogen gas explosions caused the containment vessel to release clouds of radioactive isotopes into the atmosphere affecting over two hundred thousand people living downwind from the plant. At the same time, the melting reactor core and cracks in the bottom of the containment vessel released thousands of tons of radioactive contaminated water into the groundwater table below which eventually seeped into the ocean.

TEPCO, the owner of this nuclear plant, continues to pump thousands of tons of sea water into the reactor buildings to prevent a further meltdown of the damaged reactor cores. These reactors are still far too radioactive to allow workers and equipment to enter and safely contain the highly radioactive material. The continued discharge of radioactive water into the sea from Fukushima continues even today, and the long-term effects to sea life and the increased radioactive levels observed as far away as Alaska and the West Coast of the United States will have adverse effects on sea life far into our future.

A grid down event in the United States as a result of either an EMP attack or solar storm could cause all nuclear power plants to shut down even if not initially damaged. To do a black start of all of these plants without a working electric grid will be almost impossible. Powering their computerized control systems and cooling pumps using backup diesel generators for extended down time will also be very difficult due to there being limited fuel storage on site. In addition, there is no guarantee these generators were not also damaged by the same EMP. The 2018 EMP task force report stated:

> Even totally shutdown nuclear plants must still receive backup power from the grid to keep cooling pumps operating twenty-four hours per day. In a worse-case scenario, all nuclear reactors within an affected region could be impacted simultaneously. In the United States, this would risk meltdowns at fifty-eight plant locations having ninety-six nuclear reactors, with more than 60,000 tons of spent nuclear fuel held in storage pools. Prolonged loss of power to these critical sites poses a risk of radioactive contamination to the entire continental United States.[4]

These nuclear power plants are located in thirty different states, and over 120 million Americans, or approximately one-third of us, live less than fifty miles from a nuclear plant. In addition to this potential radiation risk from these nuclear plants failing after an EMP attack, there are 2.3 million adults currently housed in federal and state prisons located throughout the United States. If there is a total loss of power and communications to a major section of our country, it's likely there will be a significant reduction in resources and staffing to keep everyone safely contained.[5]

There are currently no new nuclear power plants, let alone any coal-fired plants being designed or under construction, yet it's these simpler coal-fired power plants that traditionally kept the grid operating to black start all of these nuclear power plants. In addition, 20 percent of America's nuclear power plants are over 65-years old and designed with 1950s technology and construction materials.[6]

EMP IMPACT ON OUR MILITARY

This does not bode well for these plants surviving a catastrophic EMP event.

After several failed attempts, in December 2012 North Korea successfully placed into a high altitude orbit their Kwangmyŏngsŏng 3-2 (KSM-3) satellite followed by their KSM-4 satellite in February 2017 in a three-hundred-mile high orbit. Is it just a coincidence both satellites pass directly over the central United States as they circle the earth every ninety minutes? Prior to these successful launches, for years North Korea had been developing and testing a nuclear bomb.

During this time our nuclear scientists scoffed at the low yield of these North Korean nuclear bomb tests which were measured in the kiloton range (equal to one thousand tons of TNT). In comparison, the nuclear bombs stored in American and Russian nuclear stockpiles were measured in the megaton range (1 million tons of TNT). What our scientists failed to take into consideration is the smaller size and lower yields of the North Korean tests were an ideal size for a nuclear detonation designed to maximize EMP generation, yet small enough to fit inside a satellite placed in earth's orbit!

It's also interesting to note the North Korean KSM-3 and KSM-4 satellites are the only satellites having orbits that cross over the United States in a south to north direction, as all our military's missile and satellite early warning systems are located along our Canadian border and in Alaska. These early warning systems were designed to detect a missile attack from the Soviet Union traveling north to south, not an attack from the Gulf of Mexico! The United States has no missile early warning system at all facing to our south. In 2004, two retired Russian generals testified to the EMP Commission that the designs for their super-EMP nuclear weapon were "accidently" given to North Korea by several Russian scientists assisting with their early nuclear program.[7]

In April 2020 as this book goes to press Iran's Islamic Revolutionary Guard Corps (IRGC), internationally recognized as a state-sponsored terrorist group, successfully launched a military satellite they call "Nour," which means "light". It's 260 mile high orbit joins the two North Korean satellites as all three pass directly over the central U.S. in a south to north orbit multiple times each day, and any one of these could have an EMP generating nuclear device on board.

Russia and China are far ahead of the United States in planning the military use of an EMP weapon. The classic Russian military textbook *No Contact Wars* and the China Liberation Army's textbook *World War, the Third World War* both describe in detail the use of computer viruses and a nuclear generated electromagnetic pulse (EMP) as their primary focus in planning the next world war.[8] The military textbook *Passive Defense* by the Islamic Republic of Iran notes the decisive effects of a nuclear EMP attack to defeat an adversary.[9]

In his December 2005 speech to his top generals, China's Comrade General Chi Haotian stated:

> Only countries like the United States, Canada, and Australia have the vast land to serve our need for mass colonization. It is historical destiny that China and the United States will come into unavoidable confrontation on a narrow path and fight. Only by using non-destructive weapons that can kill many people will we be able to reserve America for ourselves. We have not been idle; in the past years we have seized the opportunity to master weapons of this kind. It is indeed brutal to kill one or two-hundred million Americans, but that is the only path that will secure a Chinese century.[10]

EMP IMPACT ON OUR MILITARY

Let me translate this for you. What non-destructive weapon is he talking about? What weapon can cause the death of up to 200 million Americans (through starvation) without damaging any infrastructure?

There are multiple countries that have both nuclear weapons and rockets capable of delivering these weapons to a specific location thousands of miles away, but our ground and satellite-based missile tracking capability would identify the launch in just seconds and then track back to where it was launched.

No doubt the country of origin would expect to receive a massive nuclear response. But what if a small group of terrorists, with or without the secret backing of another country, sailed an aging freighter near the coastline of the Gulf of Mexico or along our Atlantic seaboard. This freighter would be just one of the many ships carrying thousands of shipping containers holding tons of products and raw materials in and out of our many seaports every day. What if this small group had a camouflaged SCUD missile launcher parked under its deck or even camouflaged in a shipping container on deck?

In 1998 our military was tracking the position and radio communications from an Iranian freighter anchored in the Caspian Sea. This monitoring found the Iranian military had just raised and fired a SCUD missile off the deck using a motorized launcher. By tracking the missile, we identified it had landed in the ocean very close to a number of Iranian ships, obviously positioned to also monitor the missile test.[11] At that time SCUD missiles were still fairly unsophisticated with limited range and had poor accuracy. However, the older SCUD missile design has undergone multiple improvements since 1998 including much longer range and far better target accuracy.

A photo taken from a 2018 report by the Electromagnetic Defense Task Force clearly shows an aerial view of a stack of forty-foot-long shipping containers on the deck of a freighter. The top shipping container is clearly shown with its metal top raised up to

EMP IMPACT ON OUR MILITARY

expose a Russian missile having a fifteen hundred-mile range mounted on the launcher inside.[12] During the Gulf War there were many highly mobile SCUD missiles and trailer-mounted launchers playing cat and mouse with our military, and many launched their missiles with only a handful of crew.

These crude missiles are not accurate and most never came even close to their intended targets, but they were still a rocket that could carry a payload over a hundred miles straight up into the upper atmosphere. Keep in mind accuracy is not critical due to the large EMP effect that would be produced if the conventual warhead was replaced with a small nuclear warhead, even if the ideal 250-mile altitude is not reached. A complex and sophisticated guidance system is not needed if the rocket is launched near our coastline as it basically just needs to go straight up!

So, what if this happens? Just float along outside our patrolled thirteen-mile limit, pull off the cover, raise the launcher and fire. Who cares about aiming? Just get it high enough above our country's midsection and take down every form of communications, banking, electrical power, and transportation systems in at least half of the country. While most of our critical military communications equipment are hardened against multiple forms of electromagnetic radiation, our government uses nonmilitary commercial cell phone, e-mail, and land line services for over 95 percent of all military and government communication.[13]

As we all try to use our non-working cell phones, or impatiently sitting in cars that won't start, who do we retaliate against? For this type of attack, there will not be a long-range intercontinental missile being tracked for thirty minutes while warnings are issued, and the source of the launch is pinpointed. In fact, the ship and mobile launcher will probably be gone or sunk long before we even know the general direction the missile came from. Since the near vertical

EMP IMPACT ON OUR MILITARY

flight path lasted only seconds, it will most likely not be noticed until after the detonation.

Assuming we can identify the general location of the ship and dispatch the Coast Guard, and assuming they can select the specific ship out of the many others in the area, and assuming the Coast Guard's communications were not disabled by the EMP, odds are the small crew has abandoned ship and are now in a highspeed motorboat after scuttling the derelict ship which is now sinking. Once inside the thirteen mile limit, their motorboat or fishing trawler will be next to impossible to identify from the hundreds of other small boats in the same area.

So now what? Nobody was killed and no buildings were destroyed in the EMP attack, so do we nuke an entire country without regard to their civilian population? It's doubtful responding in kind would be very effective, assuming we could determine it was state sponsored, as many developing countries have a more rural population with limited need for computers, cell phones, ATM machines, or highways full of modern cars and trucks. And what about the rest of the world? What would the world think if the United States started bombing another country back to the stone age with a nuclear response? It appears there is not much we could do, and since many countries are less developed than the United States, many will view our crisis as their opportunity to finally achieve a level playing field with us, while we spend the next ten years trying to recover.

We know since the 1980s the United States and other major countries had large research facilities to test the effects of EMP on hardened military equipment. In addition, this testing was not limited to small military equipment. In 1980, the United States Air Force built the Trestle test center at Kirkland Air Force Base located in New Mexico that could test the damaging effects of an EMP on a full-size airplane. The United States Navy built the Empress I and Empress II test facilities at the Point Patience Navel Center on the Patuxent River in Maryland that could test the effects of EMP on almost any size ship in our fleet. Surprisingly, both the Trestle and the Empress test facilities were dismantled, and their equipment scrapped years ago.

There are smaller commercial testing centers located in several states that provide testing to certify military and commercial electronic devices will withstand lightning strikes, radio interface (RFI), high intensity radiation fields (HIRF), a solar coronal mass ejection (CME) or solar storm, and yes, even an electromagnetic pulse (EMP). Some of these commercially operated facilities have test chambers large enough to test automobiles and trucks.

If we can generate these destructive EMP waves and other forms of electromagnetic energy as part of everyday testing operations at both military and civilian facilities without detonating a nuclear bomb, then it stands to reason there are other ways to generate an EMP shockwave. There is clear evidence this research is being carried out by the United States, Iran, Russia, China, India, and North Korea, as they all attempt to create a high yielding EMP device without using a nuclear explosion. There is now ample evidence these weapons can produce a high-energy EMP pulse without using a nuclear bomb, although their destructive range is limited.

The internet is full of designs for do-it-yourself EMP weapons. Most are based on a tube that is surrounded by an energized electric coil. The tube is filled with a high explosive and as the tube expands from end-to-end during the explosion, an electromagnetic wave is generated down the coil. All you need is a small garage workshop, some basic metal working skills, and oh yes, a supply of military grade explosives! My discussions with experts in this field with years of nuclear testing experience point out these designs are like the do-it-yourself plans that claim you can drive your car hundreds of miles on a gallon of water. In other words, not as easy as it sounds, but still may be easier than we are led to believe as several recent examples have shown. It would take some exotic materials and precision machinery, but all of the materials and tools needed to build a nonnuclear EMP weapon are readily available.

A device could be loaded into a small truck or drone plane that would create a non-nuclear generated EMP wave that could destroy or disable the electrical power, computer networks, and communication systems of a small city, military base, Federal Reserve bank, or perhaps a large hospital. There would be no civilian causalities, no widespread building damage, and no nuclear radioactive fallout. The higher the level of electronic sophistication, the higher the potential for damage from an EMP attack. Since it causes no immediate loss of life, this is a weapon that is just too good not to develop with the demand for more "civilian friendly" wars. Of course, there would be extensive loss of life after an EMP attack due to the chaos it would cause to society.

While our military has been testing and specifying hardened equipment that will withstand a wide range of environmental and electronic hazards, including EMP, what about all the civilian systems that make this country function on a day-to-day basis? What is being done to keep these systems operating in the event of a nationwide EMP terrorist attack? It may be comforting for an army general to know his military-issued walkie-talkie has been tested to withstand all levels of an EMP, but what about the rest of us?

Unless you live under a rock, you cannot make it through a single day without using a cell phone, sending e-mails, making a credit card purchase, driving a car or riding in mass transit, using an elevator, turning on a light, watching television, listening to the radio, using a medical device, or heating food in a microwave oven. What if all this just stopped one sunny afternoon without any notice? Wherever you are, it all just stopped. Unfortunately, that future possibility may be closer than we realize.

In the October 2017 call-to-action, Frank Gaffney, President of the Center for Security Policy stated, "The vulnerability of America's electric grid is a ticking time bomb. The government knows that if that vulnerability is exploited by enemies or afflicted by space weather, we could experience the end of our Nation as we know it. Many of our foes are aware of both the grid's susceptibility to attack and the potentially catastrophic consequences for this country and its people should it happen. . . . Only the public is still largely in the dark about these dangers."[14]

How true that is!

CHAPTER 8

EMP Effects on Communications

Nobody really knows the full impact an EMP attack will have on our nation's communication systems or how long it will take to fully recover. We do know from actual nuclear bomb testing carried out in the 1960s that the longer the wire, the better it "couples" and will be energized by an EMP energy field. This means long phone lines, Ethernet cables, cable TV, and power lines extending from town to town and cross country will definitely receive very high voltage and current surges that will enter any electrical or electronic equipment attached at each end.[1]

Tests have shown immediately after a nuclear EMP, all communications below 100 MHz will stop, at least while the ionosphere remains energized. This would include all AM, FM, VHF, and lower ham bands. However, 2-meter ham radio, short range walkie-talkies, and cell phones all operate at much higher frequencies with less disruption, assuming the cell towers themselves were not damaged.

We also know from recent testing with EMP simulators the effects on multiple desktop computers connected to each other and to routers using Ethernet cables. The E1 pulse did burn up *all* of the Ethernet input cards located inside each of these routers and desktop computers from the radiated EMP energy. However, while the computer networks were operational once these cards were replaced, this is still no guarantee these devices will not suffer additional

EMP EFFECTS ON COMMUNICATION

damage to their power supplies from a high-voltage surge entering on the power line, or that grid power will even be available to power them.

EMP simulator testing of cell phone towers found that most did suffer damage to their electronic components and required replacement of at least some components before they could be put back into operation. In fact, EMP simulator test results are fairly consistent when it comes to microelectronic and semiconductor devices.

I have tested multiple cell phones and other handheld battery-operated electronic devices in an EMP test chamber, and they all easily survived. However, if a battery-powered cell phone or radio is attached to several feet of wire connected to earplugs or a charger, even if the charger is not plugged into a wall outlet, they will be damaged by an EMP. The wavelengths in the lower frequency range generated by an EMP are still too long to "couple" with small electronic devices, as long as there are no antennas or external wires over eighteen inches long to "collect" this incoming electromagnetic wave and send thousands of volts into the tiny microchips inside, which will be instantly destroyed.

Fig. 8-1. Typical battery-powered devices too small to suffer EMP damage.

EMP EFFECTS ON COMMUNICATION

How useful will a cell phone or radio be if there is no grid power to recharge them each night, or the cellular towers and radio stations scattered across the country are damaged by an EMP? While some cell towers do include a backup generator, most fuel supplies are only intended to last a few hours, or perhaps a few days in more populated locations. In addition, there is no guarantee the cell towers will not be damaged by an EMP, although most do include a fairly robust grounding system for lightning and high-voltage surge protection.

The internet was originally designed by military planners as a way to maintain communications if any one communication link is lost. This self-correcting system was designed to automatically reroute data packets to their end destination using any number of communication lines and internet routers. But all this data traffic assumes these hundreds of thousands of routers, data storage centers, and network computer systems have a constant source of clean electric power and the Ethernet communication links are not damaged by an EMP. In chapter 10 I will address the destruction to our banking, credit card processing, and money transfer systems if all these data communications are disrupted by an EMP or solar storm.

A large segment of today's phone and computer data communications utilize stationary orbit satellites positioned above the central United States. There are over one thousand communication satellites including, GPS positioning satellites, satellite TV, weather monitoring satellites, earth mapping satellites, and both secret military and civilian camera satellites looking down on the earth's surface twenty-four hours per day. Most of these satellites will definitely experience damage to their stabilizing systems and they could shift out of position or even fall to earth if impacted by a solar storm or EMP. It's also possible small fragments from a nuclear explosion at high-altitude could shred a large number of satellites and their solar power arrays if positioned near the

EMP EFFECTS ON COMMUNICATION

detonation point, which could completely destroy all of their power and communication systems.

After any major disruption to a city or state, people want to know what happened, is it over, who caused it, do they need to evacuate? This means emergency call centers will become flooded almost immediately for information, in addition to an expected upsurge in emergency medical, and 911 police calls. Under normal conditions these call centers handle over 200 million separate emergency calls each year and direct the action of 600,000 police and sheriff deputies, over 1,000,000 firefighters, and 170,000 medical technicians.[2]

Now imagine if none of these emergency calls are received, and all of those professional and volunteer emergency workers do not know where to go. While most of today's county emergency communication centers have multiple ways to communicate including cell phones, satellite phones, paging systems, and even 2-meter ham radio volunteers, it's still very likely one or more, or even all of these systems might not be functional after an EMP event, at least for the time it takes technicians to replace failed equipment.

A solar storm can do all of the same damage that the E3 phase of an EMP will cause, although it will not include the very fast E1 and E2 phases. In most cases there will be from a half-hour to over a day advance warning that a solar storm is about to hit, giving satellite and other communication system operators time to prepare.

I suggest keeping an emergency weather alert radio at home turned on at all times. Unfortunately, these broadcast channels are not accessible using a standard AM/FM radio, and they are quiet until an alert tone is transmitted. Every county in the United States has its own unique identification code, so you can select which county alerts you want to receive. Advance warnings of a pending solar event are also issued in addition to approaching storms, floods, and forest fires. I own a handheld emergency alert radio having a

EMP EFFECTS ON COMMUNICATION

very long-lasting battery which is always charged as I keep it in a charging stand near the front door.

In December 2018, the President's National Infrastructure Advisory Council (NIAC) published an in-depth study listing risks from both cyber and infrastructure attacks including their recommendations to reduce these risks. One of their main findings was while there were backup generators for critical facilities, there was very limited backup capability for critical communication systems. In addition to recommending ways for making traditional communication and internet data systems more resilient, the report also stressed the importance of amateur ham radio for backup to both government and utility emergency operations.[3]

Unfortunately, the state of California must not have received the memo as it started the process in 2019 to remove all ham radio repeaters used by the state's amateur ham radio operators to extend the range of their portable radio units, especially in congested cities and hilly suburbs. It should be noted that these repeaters were both supplied and are maintained by local ham radio enthusiasts at no cost to the state. The California Department of Forestry has also demanded the ham radio community start removing all remote mountaintop repeaters throughout the state or start paying thousands of dollars per year in a rental fee plus an initial $2,500 filing fee per each repeater.

The state claims their new emergency communication system provides all needed emergency communications, yet during the most recent California fires, many counties lost all cellular and land line communications, so their residents were unable to contact this new super great emergency call center. Time and again residents relied on those neighbors who had ham radios to provide all news updates regarding where the fires were headed, what areas needed to evacuate, and to report their own fire outbreaks.[4] Without these repeaters, all local ham radio operators typically operating on the

EMP EFFECTS ON COMMUNICATION

higher frequency 2-meter ham band will have their transmission range reduced to only a few miles. These repeaters allow a ham operator to communicate over hundreds of square miles with just a small handheld 2-meter transceiver.

Once these existing repeaters are removed by the state, it will be almost impossible to reinstall these later due to the astronomical costs each local group of amateur radio volunteers is being asked to pay. Again, there is no cost to the state to leave these repeaters in place, and the radio clubs do not charge anyone for their use. After the widespread 2019 forest fires in California, primarily due to the state no longer allowing yearly controlled burns to remove dead underbrush, the same ham radio groups set up their own Action Fire Net, an alert system to relay fire calls to the state's 9-1-1 system from areas where the fires had already damaged all other communication systems. Just another example of arbitrary decisions by low-level bureaucrats and state officials looking for more sources of easy tax revenue without any thought to the long-term consequences to the shortsighted decisions.

Some think the California state government views these amateur volunteer radio networks as a threat to the state's control of all communications, and removing these repeaters is just another way to keep its citizens totally dependent on the benevolence of state officials looking down from above. With the threat of a massive failure to our communication systems from both an EMP or solar storm, this is no time to tax or legislate out of existence this valuable and free emergency backup to the nation's communication systems. Let's hope the other states do not look to California and think they should do the same!

EMP EFFECTS ON COMMUNICATION

Fig. 8-2. Every June ham radio operators go portable for their volunteer field day event to see how many radio contacts they can log while operating on backup power.

Every fourth full weekend in June, more than forty thousand amateur ham radio operators across the entire nation set up tents and antennas in remote fields and parking lots for a twenty-four-hourlong competition to see how many other ham operators each can log on this weekend. This and other state events held each year allow volunteers time to practice emergency communication skills while operating on backup power. Allowing states to tax and heavily regulate this totally voluntary organization out of existence should not stand.

In later chapters I will provide several ways to keep your communication and antenna systems operating long after the grid and any backup generator have failed.

CHAPTER 9
EMP Impact on SCADA

With the nineteenth-century development of central heating systems powered by steam boilers, it didn't take engineers long to recognize the need for an automatic mechanical valve to regulate boiler pressure and the flow of steam through the piping to each area or room. With the introduction of steam powered engines, mechanical speed regulators using rotating weights maintained a constant RPM, regardless of load and without any human interaction. When centralized heating, ventilation, and air conditioning (HVAC) systems were first installed in multi-story offices, schools, and apartment buildings, these required a more sophisticated way to adjust temperature and ventilation air going to each room and each floor, so pneumatic-operated temperature controls were developed.

Using small pneumatic tubes, mechanical relays, and air dampers, it was possible to achieve much more complex control interaction, such as closing a mechanical heating valve in one location, based on a temperature sensor located in another location, all without any electrical devices or wires. While these early pneumatic control systems required yearly calibration, they were basically indestructible and totally immune to any lightning or radio interference, since no wiring or electrical devices were used. With the development of electric-powered control devices and electronic circuit boards, these pneumatic control systems were gradually replaced by more sophisticated automated control systems.

EMP IMPACT ON SCADA

Automation control systems were not just needed to control the temperature in buildings. Industry required more sophisticated controls to sequence the operation of complex machinery and control assembly lines. Soon these basic electrical control devices and banks of mechanical relays were replaced by much more sophisticated Direct Digital Control (DDC) systems that combined the computer with these mechanical and electrical control devices. By the 1990s, controlling large motors, boilers, fans, chillers, and mechanical equipment only required changing a few lines of computer programming or entering a new stop and start time with a click of a mouse.

Now a building operator could sit at their computer terminal and monitor and regulate the temperatures throughout not just a single building but an entire campus of multiple university buildings. Assembly lines and manufacturing processes could now quickly change production machinery using software to accommodate model year changes, changes to chemical formulations, and even modifying the patterns in weaving dress materials. Of course with added complexity also comes added problems. There were multiple brands of these control systems, and each manufacturer utilized their own programming language and interconnect wiring.

After these very expensive automation systems were installed, the building owner or assembly line supervisor needed to attend a distant training center run by each control manufacturer to learn how to operate their new control systems. The owners of these sophisticated control systems soon learned the only control devices that would work with their new systems had to come from the same manufacturer. Repair parts, software upgrades, and system maintenance costs saw substantial cost increases each year, and there was no way for the owners to switch system manufacturers without totally replacing their entire control system.

EMP IMPACT ON SCADA

In many cases, even the type, number, and size of the miles of wires and communication cables routed throughout their buildings or assembly plants to connect these separate control devices back to a central computer were not compatible between manufacturers and would need to be replaced if the owner changed manufacturers.

After years of price gouging and limited competition, end users and government regulators finally rebelled and demanded the control industry develop totally interchangeable control devices, standardize the communication cabling between these various computerized devices, and simplify the software and computer systems that actually provide the day-to-day control functions.

Although some minor customized features were still allowed between control manufacturers, system owners could now purchase replacement parts from almost any control manufacturer and they would work with the computerized controls and software from any other control manufacture, and "SCADA" was born.

The development of the Supervisory Control and Data Acquisition (SCADA) standards finally allowed any computer, remote terminal, microprocessor, sensor, and metering device to communicate with each other, regardless of hardware manufacturer or software developer. The control devices operating a valve, starting a motor, or switching an electric circuit could not only be done remotely using computer software, but in many cases the communications between these remote control devices and the operator could actually travel over the power wiring supplying the device, by phone line, Ethernet link, cell phone, or even satellite technology.

With the development of microelectronics, many of the smaller devices being controlled could now include their own set of built-in SCADA programming instructions, and once programmed for a specific function, they could be left alone to happily carry out their day-to-day functions without further interaction with the central

EMP IMPACT ON SCADA

control system. Today there are millions of these self-contained programmable logic controllers (PLC) operating throughout all industries on SCADA networks and they are extremely durable and reliable.

Fig. 9-1. Programmable logic controller (PLC) for SCADA networks.

These PLC hardware devices carry out the SCADA software commands to control almost every remote acting valve, fan, motor, and pump in today's modern water treatment plants, sewage treatment plants, electrical sub-stations, power plants, assembly lines, oil refiners, gas transmission lines, lift gates on dams, elevators in high-rise buildings, railroad routing, subway stations, and even the traffic light on your street. If you work in an office building or attend classes in a modern school facility, it is an almost 100 percent certainty the thermostat on the wall, the air flow from the cooling vents, and even the classroom lighting are all now controlled by these small PLC devices interconnected with other devices in the system and back to a central computer console utilizing SCADA control programming.

EMP IMPACT ON SCADA

Unfortunately, while SCADA and PLC devices are very reliable, they still require a working electric grid to operate. In addition, the tiny microchips embedded in each PLC and SCADA control device will be "fried" by an EMP-induced voltage surge. Even if backup power is available, or grid power can be temporarily restored, the loss of these reliable control systems still means many of the services, utilities, and processes we take for granted may still not function.[1] Even worse, it's possible if grid power can magically be restored soon after an EMP grid down event, the processes and systems the failed devices controlled may just sit there not knowing what to do, or worse, operate in an out-of-control nightmare.

Fig. 9-2. SCADA control boards "fried" during EMP testing.

Loss of power to SCADA control systems is not the only concern. During EMP testing, computer networks experienced corruption and changes to the data being sent between the separate desktop computers. This does require the computer devices to remain powered and operational, but this is still possible with a working backup power system. However, the potential for data corruption to the bits and bytes traveling over the communication networks of SCADA control systems still operational is a real concern.

EMP IMPACT ON SCADA

Lift gates on dams might lower on their own, elevators may decide not to stop as they speed to the ground, automated mixing of chemicals in water treatment plants could change or perhaps not provide any treatment at all. SCADA controls and computer systems monitor the position and sequence thousands of subway trains in a city, and routing hundreds of freight trains crisscrossing this country. What if these automated SCADA controls dispatching and switching tracks for thousands of tons of freight and passenger cars suddenly decided to stop working or fail to switch tracks when directed by the central system?

The San Diego County Water Authority uses SCADA controls to remotely operate massive motorized valves that supply over 825 million gallons per day of water passing through their aqueducts. Any disruption in the remote control of these constantly changing valves could cause a catastrophic failure of the piping system and massive flooding of private property. The San Diego Gas Company also uses a SCADA-based automation system to remotely control hundreds of large gas valves located throughout their distribution system to maintain the proper flow of gas and to prevent a blowout.[2]

In November 1999, both of these California companies suddenly found they had lost all remote control of all valves from their central operating centers and had to dispatch emergency work crews to remote areas of the large county to manually close those large valves. Although these work crews reached these valves and were able to shut them down in time, having both systems fail at the same time could have caused unimaginable property damage and loss of life. After both systems were brought back to temporary operation with manual control, it was discovered their wireless SCADA control systems were being interfered with by the radar system on a Navy ship located over twenty-five miles away off the coast of San Diego!

As will be addressed in chapter 11, the first-generation of electronic ignition systems introduced on new car models in the early

EMP IMPACT ON SCADA

1970s, would stop the car engine when driving past an airport each time the tower's rotating radar dish pointed in their direction. Again, this begs the question: if the low level of electromagnetic radiation from a radar dish twenty-five miles away can stop cars and take down the entire water and gas distribution control system for San Diego County, what kind of havoc can a much higher level of EMP energy cause?

EMP simulators have demonstrated how PLC controllers may continue to operate if power is available but suddenly send erroneous data back to their SCADA central control systems. What if an EMP caused the water level sensors in a nuclear power plant to show the water level was full when it was actually low (China Syndrome). What if combustion sensors sent the wrong air-fuel mixture signals to the large motorized fuel valves on a multi-story coal-fired boiler?

Many of these automated SCADA control and monitoring systems are located in very remote locations and use cellular or internet communications to interconnect and coordinate with their central control centers. Tests have shown an EMP can alter digital data being sent over Ethernet cabling or even totally destroy control sensors.

A follow-up test was conducted by the North American Electric Regulatory Corporation (NERC), using desktop computers connected through an Ethernet router and standard Ethernet cables, which were subjected to EMP from a portable EMP simulator. The first test positioned all four computers twenty-five feet apart; the second test moved them to two hundred feet apart. Test measurements found the EMP currents generated in the two hundred-foot ethernet produced by the portable EMP generator were two times greater than when they were connected with the twenty-five-foot cable, and current peaks varied between 100 and 700 amps! Needless to say, all computers and routers suffered failure to their communication hardware.[3]

EMP IMPACT ON SCADA

In the late 1980s, there was a catastrophic failure of a thirty-six-inch gas pipeline just one mile from the city of Den Helder in the Netherlands. Although the resulting gas explosion caused extensive property damage, it could have caused a major loss of life if it occurred just one mile closer to the city. The failure was caused by a remotely controlled gas valve opening and closing each time the radar dish at the nearby airport rotated around and swept past the SCADA device controlling this large valve. This rapid cycling of this valve caused a pressure wave to travel down the long pipeline which caused the pipe rupture and explosion.[4]

Many of these networks of interconnected SCADA control computers connected to remote sensors and actuators operate totally without any human intervention, yet most of these control devices have no shielding against EMP, and their interconnect cabling will serve as perfect antennas to collect and distribute very high currents and voltages generated by an EMP. Keep this in mind the next time you ride an elevator!

Of course, not every SCADA device will fail. Some will be shielded from the EMP due to better shielding of their interconnected control wiring, their mounting orientation, or perhaps just plain luck. Until an EMP event actually occurs, there is just no way to predict what will fail, what will not fail, or which devices can be made operational again by reprogramming or a restart. This is just another EMP concern that few are even discussing.

CHAPTER 10

EMP Impact on the Banking Industry

Probably the most long-term damage to our nation from an EMP attack or solar storm will be to our banking systems. The private Federal Reserve Banking System establishes monetary policy of all member banks, determines interest rates, and extends emergency credit to its member banks. The Federal Reserve System consists of twelve district banks located in Richmond, Virginia; Kansas City, Missouri; Minneapolis, Minnesota; Chicago, Illinois; Atlanta, Georgia; St. Louis, Missouri; Dallas, Texas; San Francisco, California; Boston, Massachusetts; New York City, New York; Philadelphia, Pennsylvania; and Cleveland, Ohio.

Below the Federal Reserve are commercial banks, brokerage houses, and mutual fund traders. Today there are almost no paper transactions, so *all* transfers and verification of cash and credit flows between these banks and their clients and customers are now by electronic fund transfer. All evidence of each transfer is just the bits and bytes traveling over wires and stored on magnetic tapes and disks.

The FEDNEC data network connects only the twelve Federal Reserve Banks, while the FEDWIRE data network connects over 7,500 local banks and financial institutions. All fund transfers over these two electronic networks are instant and irrevocable. The average daily transfer using FEDWIRE is $2.8 trillion, and the system handles over 534,000 payments each day. Local financial institutions utilize other electronic fund transfer networks, which process over

EMP IMPACT ON THE BANKING INDUSTRY

300,000 interbank transfers per day, and millions more electronic transfers are made each day for personal check processing. The SWIFT network is a private e-mail system used between banks, brokers, and stock exchanges which relays over eight million fund transfers over the internet each day.[1]

In total these electronic communication networks handle millions and millions of financial transactions each day. In addition, all back up files for these transactions are stored electronically on magnetic disk drives and not on paper files or microfilm as was done in the past. Not only could a large area EMP damage some or all of these multiple electronic networks but could also result in data errors or destroy backup data centers leaving no written record or proof of who owns the billions of dollars that were in process just prior to an EMP attack.

Although these private data networks involve bank-to-bank and bank-to-Federal Reserve fund transfers, the most immediate impact at a consumer level will be the inability to verify personal checking account balances, being locked out of all ATM cash machines, and unable to use a credit card. Overnight we could become a cash-based society again while taking weeks or months to get the banking system operational again. Attempts to clear up the backlog of unprocessed money and securities transfers could take months. Due to the wide geographical areas involved, an EMP attack could damage our financial institutions that far exceed anything a cyber-terrorist could hope to achieve.

According to the President's National Security Telecommunications Advisory Committee, "the financial services industry has evolved to the point where it would be impossible to operate manually without information technology and networks."[2] The Federal Reserve Board noted their financial operations would be seriously impacted if their communications were disrupted for even a few minutes. They noted all inter-bank fund transfers, security transfers, and treasury note processing require same day recovery for the stability of the financial markets. A review of all past national and manmade disasters did not find where a single bank was able to reopen using paper or handwritten records during a power or communication outage, as all were forced to close immediately.

The power outage in August 2005 caused by Hurricane Katrina disrupted normal business and banking transactions for months in Louisiana and caused an enormous loss to the state's economy. Attempts to go back to paper transactions while all lost electronic records were being sorted out was plagued with fraud and human errors. All of the many safeguards for handling your money are built into the paperless electronic transaction systems and backup data files, so theft will increase significantly if our economy had to suddenly revert back to cash only after an EMP or solar storm destroys these banking communication systems and computer networks.

The entire financial system of the United States plus the fund transfers with many other countries each day are all now done by electronic communication. All records of these hourly and daily transactions are stored electronically in huge data storage centers. No doubt there are backups and even backups for the backups of all this transaction data, but everything is still stored electronically, and all risk data errors and loss of access if power is lost after an EMP.

Anyone replacing the hard disk drive in their home computer is warned multiple times to electrically ground the work surface, their tools, and both hands first, as any static charge can damage the data stored on the magnetic disks inside. If an accidental "spark" from shuffling your feet access the carpet can damage a magnetic memory device, imagine what 50,000 volts from an EMP would do that simultaneously enters each of the thousands of hard drives tirelessly spinning away in these data storage centers!

An EMP cannot erase data stored on backup magnetic tapes or CD disks. However, an EMP can damage the disc and tape drives

EMP IMPACT ON THE BANKING INDUSTRY

used to store this data, potently making it difficult or impossible to retrieve the stored data. In addition, the financial sector is constantly moving a huge volume of data between these institutions, so even a power interruption lasting only minutes cannot be allowed. For example, a single credit card company recently processed over five thousand credit card payments *every second* during this past Christmas Holiday.

History has shown what happens to other countries that suffer a major collapse to their society or a major grid down event lasting months. Once utility power is lost and all communication systems and computer networks are down, paying for anything with a check or credit card will be impossible, regardless of how much your bank's website showed you had in your account just before things crashed.

When this happens, anyone with cash in hand or stuffed in their mattress will be king and will be able to buy almost anything for drastically lower prices, especially when everyone hasn't eaten for days. However, history has also shown at some point merchants and store owners with registers stuffed full of cash and now empty store shelves will realize if there is no functioning government or banking system, all this cash will basically become just worthless green paper. They should have held on to their merchandise which could at least be used for barter.

Gold and silver coins will not only retain their value during times of economic stress, their value will skyrocket. Unfortunately, you can't buy a loaf of bread with a gold coin worth two thousand dollars if nobody can make change, so most day-to-day exchanges for food, clothing, fuel, and medicine will be either by barter or theft. During hard times, local swap meets and farmers' markets typically spring up almost overnight, and believe me, hard times are coming when, not if, our electric grid is taken down by an EMP or solar storm.

While I am not suggesting we all need to keep a few gold bars under the bed just in case, there are several low-cost things we can do. I recently drove over fifty miles to a specialty store to pick up a tool I could not find locally. After spending an hour on the road, then another half hour in the store picking out exactly what I wanted, I headed for the cashier. Noticing lots of people just standing around but not in line, I placed my selection on the counter and then was told their entire computer system was down and could not process any credit card or check payments.

Being a large chain store, their cash registers were down nationally, and it was expected to take most of the day to get things working again. No doubt somebody in their IT department is now looking for a job! Fortunately, I had some backup cash but hadn't planned to spend it that day. Having cash allowed me to make the purchase and avoid a second trip. You know, sometimes bad things just happen, and it does not take an EMP event to strand you in some remote town unable to buy gas or food while just trying to get back home.

I always considered a credit card as a way to spread out the cost of a big-ticket purchase, or perhaps cover an unexpected car repair, but still carry cash to cover my everyday type smaller purchases. Obviously, today's younger generation has totally given up on cash, which is what this government wants as cash purchases cannot be tracked and can sometimes be used to avoid paying taxes. My work requires traveling many days a month, and most lunch times will find me in a fast food line. I am absolutely amazed that most of the people in line ahead of me are all paying with a credit card, even for a four-dollar sandwich. While I suppose they could have a wad of cash in their pocket, I suspect many do not. While it's convenient to "wave" these newer type credit cards near a data terminal and no longer even need to hand the card to a clerk or sign a receipt, what happens if this credit card system quits working?

Every vehicle has an owner's manual in the glovebox, and this is a perfect place to keep enough cash in small bills to purchase at least two fill-ups of gas. This should get you home and even buy dinner, instead of being stuck in a distant town driving on fumes because a power outage or screwed-up computer network has stranded you with nothing but useless plastic in your wallet.

At the next level up, you should keep enough cash hidden somewhere at home to purchase several weeks of groceries, just in case. What if a major storm event is forecast and you rush to the store only to find its cash only? If the grid or computer networks are down for any reason, it makes no difference how much money your bank statement shows, if you can't use a credit card or money machine you are basically broke without access to cash!

Being financially prepared for an EMP or solar storm will also prepare you for any banking related crisis or even just an unexpected computer screwup.

As we close out this chapter, it's important to remember that historically many societies and countries have begun, grew, matured, stagnated, then finally collapsed back again for thousands of years due to natural catastrophes and wars. However, there were always survivors. If you choose to be a survivor, you need to prepare for a time without electricity, and when the money you had in the bank, including your retirement plan, may have just magically vanished!

CHAPTER 11

EMP Impact on Vehicles

In researching this book, I met with firms that operate large RFI and EMP simulators used to test the "hardening" of military and aviation equipment against damage from all forms of electromagnetic energy. Prior to the early 1970s, vehicle ignition systems were basically a mechanically-driven distributor, cam lifted points, and an ignition coil. Most engine gauges and the speedometer were also nonelectric mechanical devices which almost guaranteed there would be no engine damage from an EMP. However, it is still possible for these older vehicles to experience damage to their lighting systems, radios, and other electrical accessories.

Most modern vehicles have much better radio frequency interference (RFI) shielding due to the extensive use of computer devices and a digital dashboard display. It's possible that many vehicles will survive an EMP attack with only minor damage, but this still does not mean there will not be a major disruption to our modern transportation systems. Even if your own vehicle is not damaged, there is still the problems of blocked roads by abandoned vehicles that did suffer ignition damage or ran out of fuel, major traffic disruption due to non-operating stop lights at every intersection, and closed service stations and vehicle repair shops due to lack of grid power.

Actual EMP testing of vehicles has shown most will initially shut off and coast to a stop when hit by an EMP if the engine was operating at the time. In some cases, these tests found if the car's battery was disconnected completely for a short period and then reconnected, this would reset the vehicle's computer devices which would allow the engine to be restarted for some of the vehicles tested. This is similar to turning a desktop computer off and then back on to reset once it is locked up and displays the "blue screen of death."

Older diesel engine–driven generators and heavy construction equipment having a mechanical-driven fuel pump and mechanical driven fuel injectors operate with no electrical ignition system at all. I have witnessed a diesel dump truck continue to run for twenty minutes while totally upside down at the bottom of a ravine after a wreck while serving as a county EMT!

Just like today's gasoline car engines, most new trucks with fuel-efficient diesel engines have converted over to all electronic systems to control the metering and distribution of diesel fuel to each cylinder, digital controls to monitor the position of the accelerator pedal, and a digital dashboard display. These newer diesel cars and trucks will have the same type problems as gasoline vehicles when exposed to high levels of electromagnetic energy.

A modern car or truck can have as many as one hundred microprocessor-controlled devices and electronic sensors interconnected to literally control everything. Stop and think about how a modern vehicle is full of electrically powered devices controlled by the vehicle's multiple microprocessors. This includes the air conditioner, airbag deployment, and backup camera, anti-lock braking, automatic transmission shifting, battery, CD player, collision avoidance sensors, cruise control, digital dashboard display, directional lights, electric door locks, electronic odometer, electronic speedometer, hazard lights, headlights, heater, horn, interior lights, multi-speaker sound system, navigational GPS, outside temperature, power steering, powered windows, radio, seat belt alarm, seat position, suspension adjustment, tire pressure, and windshield wipers. There are also the actual engine operating controls including digitally controlled fuel intake, electric choke, electronic ignition, tachometer, temperature, charging alternator, fuel efficiency reporting, and exhaust emission control.

Many of these controls utilize miniature microchips and integrated electronic circuits which are easily damaged by high-voltage and current surges. In addition, most interior and exterior light bulbs and instrument lights now use LED type lamps, and testing has shown LED lamps will not survive an EMP if connected to a long wire.

To even further highlight the amount of sensitive microelectronic devices in new vehicle technology, the transition over to hybrid and all-electric vehicles is adding an additional layer of electronics to control energy flows into and out of a large battery pack that is not found in a convenutal vehicle's electrical system, and I won't even begin to list the highly sophisticated computer controls with millions of lines of computer code and multiple camera and position sensors required to make self-driving vehicles a reality. So, will our vehicles withstand a real EMP event or not? The answer is nobody knows for sure and luck may play a bigger part than first thought.

There is also a question regarding how an EMP will affect the newest battery technology in use for the all-electric vehicles. The popular Tesla model S all-electric cars have sixteen battery packs under the floor, and each pack contains approximately five hundred batteries. Each of the batteries is similar in size to a standard AA flashlight battery, except they are slightly larger in diameter and one-third longer. In addition, these battery packs are extremely sensitive to low temperatures, high temperatures, and unequal charging of each battery cell, so each battery pack includes a built-in electronic battery management system (BMS) that keeps charging, discharging, and temperature limits in proper balance.

EMP IMPACT ON VEHICLES

This means a typical all-electric car may have up to eight thousand individual batteries, and a separate computerized battery management system for each battery pack. The Chevy Volt and Nissim Leaf also include multiple battery packs with similar numbers of individual batteries and a microprocessor based BMS system for each pack or tray.

While it is expected an EMP will probably not damage the eight thousand separate batteries, there is the risk of arcing inside these battery packs, and the potential failure of a number of the sixteen thousand separate tiny battery connections which will act like fuses under high current flow. Tests have shown conventional liquid filled, Gel, and AGM lead-acid batteries are not affected by an EMP or solar storm since they do not contain a built-in microprocessor controlled BMS to monitor the charging and temperature of each individual cell of the battery. All-electric vehicles typically require a working electric grid to recharge each night, so many will be sitting motionless in their garages during a major grid down event.

The Congressional EMP Commission's final report submitted in April 2008, indicated thirty-seven cars and eighteen trucks were tested in full-size EMP simulators. These vehicles were randomly selected from model years between 1986 and 2002 and tested with both their engines running and engines not running during the tests. The intensity of the EMP energy field varied from 10 kV/m to 50 kV/m, with the higher value representing the level of EMP energy a high-altitude nuclear explosion is expected to produce. Many automobiles did not have any adverse effects at half the anticipated EMP levels, but 10 percent experienced serious problems at the maximum EMP testing levels. This included engine shutdown, dashboard display errors, or accessory control problems.

Most problems occurred to cars with engines running during the tests, yet eight cars experienced no problems and continued to operate normally even at the highest levels of the EMP test. The trucks tested included both gasoline and diesel engines, and

EMP IMPACT ON VEHICLES

everything from pickups to tractor trailers. The trucks with engines not running during the testing did not experience any adverse effects, however, three trucks did have their engines stop that were running during the testing. Two trucks were easily started after being shut down during the tests, and one truck required towing to a garage. Ten trucks exhibited minor or temporary problems during the testing, but none required driver intervention.

Several vehicles with stopped engines, nonfunctioning dashboard instruments, or meters showing data errors were able to be reset back to normal after the vehicle's battery was disconnected for a short period of time. Based on this testing, it is estimated that 10 percent of all cars and 15 percent of all trucks will experience engine shutdown when exposed to the higher levels of EMP energy, but many will be repairable. In full disclosure, the vehicles used in this testing were loaned from existing government surplus inventory or sister agencies. The committee was required to stop testing at the first sign of any electrical or engine problem, as the vehicles were on loan and were intended to be returned to normal service.[1]

This means some vehicles could have suffered catastrophic damage if the tests were not stopped at the first sign of a minor problem. In addition, these tests did not include any vehicles manufactured after the 2002 model year. My only comment is I hope you are not in rush hour traffic, surrounded by thousands of other commuters barreling home, while traveling at seventy miles per hour when an EMP strikes! While you may be able to restart your car later, it's also possible you might be in the middle of a five-hundred-car traffic jam.

Chapter 31 has more information regarding simple ways to reduce your vehicle's vulnerability to an EMP attack.

CHAPTER 12

EMP Effects on Transportation Control Systems

Related to EMP effects on vehicles discussed in chapter 11, over 80 percent of all traffic signals are now microprocessor and SCADA controlled, including multiple under-pavement and camera vehicle position sensors. These computerized traffic controls were tested using moderate levels of EMP exposure by the EMP Commission. They found all traffic control programs became corrupted after EMP exposure with combinations of green, red, and yellow lights randomly turning on and off, or reverted back to their default mode of just flashing on and off.[1]

Although most of this damage could be repaired by a simple control board replacement, the sudden impact of 10 percent of all cars and 15 percent of all trucks experiencing engine failure at the same time a large number of traffic lights are randomly flashing unsynchronized colors or not operating at all due to loss of grid power could be disastrous. There will be multiple car accidents at every intersection if this happens. Assuming you are driving one of the cars or trucks that were not affected by an EMP, you still may be prevented from driving any great distance. If all gas stations are without power to pump fuel or process credit cards, and major highways and Interstates have turned into parking lots packed end to end with abandoned cars, traveling by car any real distance will be extremely risky.

EMP EFFECTS ON TRANSPORTATION

Transportation in the United States is not just our personal cars and pickup trucks. Grocery stores typically have less than a three-day supply of canned goods, fresh vegetables, meats, and all kinds of frozen foods. Many of these products are coming from producers and growers located thousands of miles away yet arrive like clockwork on a daily basis. Through a sophisticated computerized system of bar code tracking of customer purchases and "just in time delivery" to restock store shelves, it's rare the item we need is out of stock. This requires a constant stream of tractor trailer trucks, railroad boxcars, and even air freight for the most perishable products.

Every day the skies over the United States experience forty-four thousand flights carrying over 2.2 million passengers departing from 5,087 public airports.[2] The failure of the ticketing computer for a single air carrier lasting only a few hours is not an usual occurrence yet can cause a near standstill of all flights stranding passengers and taking days to sort out. An EMP or solar storm can shutdown not just one but all carrier ticketing computer systems, plus control towers, multiple navigational radio beacons, and ILS landing systems all over the country. While almost all of these systems do have emergency backup generators, it's realistic to expect an EMP will also cause failures of at least some of these backup systems, and generators do not run forever without fuel.

There is also a major concern for the airline industry, as today's large passenger planes, except for the older 747, are now fly-by-wire, meaning electronic control devices are used from end to end to electrically carry out the pilot's control movements, auto pilot, position of wing and tail controls, cabin pressure, switching between fuel tanks, and even raising and lowering landing gear. Even with an engine failure, the controls still allow the pilot to safely maneuver the plane to a landing, as emergency power is still available to operate all these control devices. But if each of the control devices and their tiny microprocessors are damaged by the high-voltage surge from an

EMP EFFECTS ON TRANSPORTATION

EMP into all of the various antennas, the aircraft would basically become a flying brick. While the metal skin of an aircraft would provide some attenuation of the EMP shockwave, all of the windows plus radio antennas sticking out in multiple directions will do just fine collecting this EMP energy and directing it into the critical electronic systems inside.

Recently a helicopter crashed after flying too close to an antenna farm of a major radio and satellite communication center. The investigation found the strong radio interference caused the immediate failure of the helicopter's computer-controlled fuel system.

Due to GPS jamming, an Embraer three hundred passenger jet dropped over fifteen thousand feet when the altitude stability computer controls lost the satellite GPS signal.[3] While this demonstrates what minor radio frequency interference (RFI) can do, a real EMP event will blanket the country with an electromagnetic shockwave millions of times stronger and containing a much wider frequency range.

As introduced in chapter 8, an EMP event will damage many of the over one thousand satellites currently circling the earth. Including in this total are thirty-two satellites and nine spares located directly over the United States that makeup the Global Positioning System (GPS) network.

Today many of us rely on a GPS dashboard display to take us effortlessly to our destination while its constantly updating traffic information when driving through an unfamiliar city. A large segment of both private and commercial aviation now uses GPS signals for both navigational position and elevation data. Many private boats and commercial shipping also use GPS signals for either primary or secondary positioning data. Commercial and military drones sometimes use GPS signals to direct them to a very specific destination, with an accuracy of less than a foot. Even if grid power

EMP EFFECTS ON TRANSPORTATION

can be restored in a reasonable time, it could take years to replace all of these satellites.

While an EMP can impact our personal vehicles as discussed in chapter 11, an EMP will also impact our nation's rail transportation system. Like the computerized traffic control centers in each large city controlling traffic lights and monitoring the flow of traffic, our country's rail system also requires multiple computerized control centers. These direct which rail cars are assembled into which specific train, and which route a given train will take once it's assembled. In addition, unlike our Interstate system of multiple traffic lanes heading in each direction, a large percentage of the railroads crisscrossing our country are just a single, or at best, two separate sets of tracks.

A single train doesn't just head out from Virginia and travel all the way to California, for example, before another train is allowed on the same track coming back in the opposite direction. There are hundreds of trains traveling in both directions twenty-four hours per day and in some cases on the same set of tracks, with fast passenger trains overtaking slower freight trains heading the same direction. During all this, freight trains constantly stop to drop off and pickup additional freight cars while passenger trains stop to pick up passengers.

This is only possible by the use of an automated system of block signals every few miles, thousands of remote-controlled track switches, and the scheduling of all locomotives by radio and satellite communications. All these separate systems must be synchronized together to make this possible.

While an EMP or a solar storm can damage all these computerized railroad dispatch centers, thousands of automated track switches, and grade crossing lights, they all still require grid power to operate, although many do have limited emergency battery backup. About twenty percent of all operating diesel locomotives do not have microprocessor devices or computer controls. However, the remaining 80 percent of all locomotives in operation today utilize sophisticated computerized controls to manage multi-wheel traction motors, braking, engine function and alarms, and provide easier operator interaction.

While these older diesel locomotives are considered to be fairly immune from EMP damage, the computers and SCADA control devices required to operate all newer locomotives are susceptible to damage from an EMP or solar storm.[4] While the all-metal locomotives are expected to provide additional EMP shielding for the sophisticated computer control systems found inside, they are still not immune. Since all locomotives are designed to come to a complete stop when there is a failure of their computer controls or an incapacitated engineer, we probably will not see a driverless locomotive racing down the tracks after an EMP attack! However, there could be a total stoppage of all rail traffic after an EMP due to this loss of controls, and it could take months to get all these trains unlocked and running again.

According to the United States Energy Infrastructure Administration (USEIA), in 2018 over 40 percent of our nation's electrical generation was still by coal-fired plants. While the use of coal is dropping in the United States due to stricter air pollution regulations and a recent drop in natural gas prices, this is still a sizable market share. While 70 percent of all coal delivered to our nation's coal-fired power plants is by rail, these power plants typically store only a thirty to sixty-day stockpile of coal on site.[5] Assuming EMP damage to the computers and SCADA devices that operate these coal plants did not stop their ability to generate power, without continued rail service, many will still power down within weeks due to the loss of an adequate fuel supply.

CHAPTER 13

EMP Impact on Oil and Gas Distribution

Almost nobody thinks about underground oil and gas pipelines since you cannot see them, they don't move, and it's rare they cause a problem. From time to time when driving through the country you might cross a long cleared path of woods approaching the highway from one side, then continuing away on the other side that looks exactly like the clearing made for high-voltage power lines, but there are no visible power lines and no electrical towers.

Occasionally you might see where a large pipe comes up out of the ground, passes through a few valves, then returns underground. This is part of the many underground gas and oil pipelines crisscrossing the United States delivering crude oil from terminals and well fields to refineries, then sending multiple grades of refined oils and gasoline to their distribution centers many states away.

There are more than fifty-five thousand miles of twenty-four-inch diameter crude oil pipelines, plus an additional forty thousand miles of six-inch diameter pipelines from oil well fields to their central gathering points. Over 50 percent of all crude oil processed in the United States travels through these pipelines. These oil pipelines are critical to move these refined oil products out to the rest of the country.[1]

The entire process of refining and piping the oil and gasoline is now highly automated and relies totally on SCADA computer controls and many thousands of interconnected PLC devices to

EMP IMPACT ON OIL AND GAS DISTRIBUTION

remotely operate valves and pumps, to monitor flows, adjust pipeline and storage tank pressures, and route these flows into and out of various piping systems. The potential failure of SCADA-controlled PLC devices has been well documented in chapter 9 and I will not repeat here. However, any damage to these thousands of SCADA computer control systems by an EMP will guarantee these piping systems will need to be shutdown manually or risk major oil spills, plant fires, and equipment damage.

There are more than five hundred natural gas processing centers in the United States, and more than fourteen hundred remote compressor stations spaced along the more than three hundred thousand miles of underground gas transmission pipelines. SCADA controls are critical to the operation of these underground natural gas pipelines.

Maintaining the correct pressures in these gas pipelines and constantly adjusting each compressor station requires a very sophisticated system of interconnected computer controls, flow sensors, and pressure sensors. In addition, these SCADA pressure and flow sensors may be located hundreds of miles away from the compressor stations they control. This demands a failsafe data communication network between the compressors and the pressure sensors to constantly provide instant feedback.[2]

Not only can an EMP cause failure of these SCADA and PLC pressure and flow controls, but an EMP can cause data errors in the communications between flow and pressure sensors and the compressor stations. This could result in an overpressure pipe rupture that can cause extensive fire and explosive damage to any structure and people nearby.

While these natural gas pipelines are usually buried over five feet deep, when there is a rupture and explosion, the deep crater and scorched earth for hundreds of feet around looks like a bomb exploded. Immediately after an EMP, the loss of grid power to

EMP IMPACT ON OIL AND GAS DISTRIBUTION

operate these large compressor pumps will cause the gas pressures in these piping systems to just suddenly drop to zero and stop. Manual intervention may allow the safe shutdown of these piping systems after the loss of grid power, but this will not allow the system to remain operational.

Although oil refineries of various sizes are located in thirty-five different states, by far the states with the most and largest refiners are Louisiana and Texas. Oil refineries are now highly automated with only a few operators needed to remotely control thousands of valves, pumps, steam boilers, and distillation columns to separate out the various petroleum products.

All of these chemical processes depend on thousands and thousands of SCADA controls and PLC devices to carry out the computerized commands from the control centers. Tests have shown all SCADA devices are susceptible to EMP damage. In addition, loss of control during the refining process can cause pipes and large storage tanks to rupture, spilling thousands of gallons of hot flammable oils causing fires and explosions. It could take months to replace all of the damaged SCADA devices before operations can resume, and this assumes there was limited damage from the fires and explosions.

Even the temporary shutdown of all oil refineries in the United States lasting only a few weeks or months would be devastating to the country's transportation systems. While oil and other fuels could be imported from other countries to supply vehicles and other industrial processes not damaged by an EMP, this still requires the pipelines to transport these fuels from tanker unloading facilities on the coast to inland distribution centers. It is expected most of these pipelines would be down at the same time after an EMP due to these SCADA control issues.

While it may be easy for some to say replacing all the damaged SCADA controls is not a big deal, but how will all the service

EMP IMPACT ON OIL AND GAS DISTRIBUTION

technicians, system operators, and parts supplies get back and forth to work for the weeks or months it will take to make all these repairs? How will all the needed parts be manufactured and delivered without grid power? What is normally a simple task when you have water, power, food, transportation, and communication systems, is not that easy without these. History has shown when there is a major disruption to any society, civil unrest is soon to follow. All those police, fire, and rescue first responders we have come to rely on will not go to work if they feel their own families are at risk and need their protection.

CHAPTER 14

See Who Is at the Door

Now that you have learned how an EMP and solar storm can impact our society and everything in it, the next part of this book will hopefully make you better prepared for whatever we all will face immediately after it happens. While I do not believe either event will cause the total collapse of our society, it definitely will cause hardship, hunger, security risk, and illness for millions, especially for those who have not made any effort to prepare.

While current disaster movies depict huge masses of starving zombies killing everybody that gets in their way, I don't believe things will get quite that bad, at least I hope not! However, I am not optimistic that our government and the utility industry will ever find the money, the common ground, or the leadership to actually come together and harden the electric grid. I think it's only a matter of time before those countries who are our sworn enemies, and perhaps terrorists supported by them, will find the opportunity to take our grid down. Afterall, look what a few men could do with nothing but box cutters in four planes full of unarmed and mostly compliant crews and passengers. It's also statistically likely that we are way past due for a major solar storm with similar results.

The most obvious and simplest causes of past major disasters started out as not that obvious until after they happened, which implies we probably won't see it coming. Barring an all-out nuclear war involving thousands of missiles raining down on us all, successfully carrying out a single nuclear EMP attack will still not be capable of destroying every single electric appliance, vehicle, and computer system across this entire nation. However, it will seriously impact a large number of our cities and citizens. Actual real-world nuclear tests have shown there will be extensive damage to the electrical grid within several hundred miles of the bomb's detonation, plus some damage to interconnected electrical systems up to a thousand miles away.

EMP testing has shown the engines in many vehicles will initially stop, but some can be made operational again, and some vehicles may not be affected at all. No doubt there will be severe damage to the many high-voltage transformers in our electric grid and the largest transformers may take over two-years to manufacture and ship. While hundreds of power plants and their grid networks have been interconnected for years, our nation's grid system is still divided up separately into the western grid, the eastern grid, and the Texas grid. The electric utility industry is currently making efforts to further subdivide these three separate networks so the failure of one grid system will no longer take down hundreds of others, as has happened in the recent past, but this is still a work in progress.

While there has been a major increase in both residential and utility-size solar power systems, without backup batteries these systems cannot produce any power if the grid is down, so utilities are looking into more use of "microgrids." This will allow smaller sections of any grid system to continue to operate independently of other parts of the grid when surrounding areas are without power. There are now over 1 million all-electric cars in the United States and growing exponentially, yet without grid power none will be running for long.

On August 14, 2002, the country's second largest electric grid blackout hit midwestern and northeastern sections of the United States and eastern Canada. Ohio's First Energy had several high-voltage transmission lines touch each other and nearby trees during high winds. When grid operators failed to take corrective action, this caused a cascading shutdown of 256 power plants across the entire east coast that were interconnected. By the end of day over 55 million people were without power in eight states, and some areas took up to a week to fully restore power. The problem was finally traced to a simple "bug" in First Energy's computerized control software which blocked an alarm from reaching the system operators.[1] While interconnecting separate grid systems into larger networks can provide added system capacity and more reliability, it also means a higher risk for the entire system going down if one part of the country is hit by an EMP.

During the solar storm of March 1989, that struck Québec and several northeastern states interconnected with Canada's hydroelectric grid, several high voltage transformers and substations were seriously damaged and six million people were without power. Power was restored to most areas within days due to the availability of emergency crews and spare parts.

This was still just a glancing blow from a small solar storm that only hit a less populated area of eastern Canada. If the earth had been positioned with the entire continental United States facing directly into the path of this solar storm just as it crossed our path, there would have been severe damage to our entire grid from coast to coast, and a full recovery could have taken years. While I hope we will not be facing the total collapse of our society, I think it's very naive and short sighted to think we will not need to worry at all.

Obviously if you are forced to evacuate and our nation's entire grid system is down for years, followed by a complete breakdown of civil society, then all bets are off. If this actually happens, then perhaps William R. Forstchen's novel *One Second After* will give you a better understanding as to what a true societal collapse can do after a grid down event. I am hopeful many of us will be able to shelter in place and still maintain an acceptable lifestyle for the months it takes to restore power in all affected areas. However, there are a few

SEE WHO IS AT THE DOOR

questions you do need to consider first, and these are difficult questions requiring some real soul searching.

This chapter will be the starting point for anyone wanting to live comfortably while staying home during a major grid down event, regardless of the cause, assuming you have followed the preparations discussed in the next few chapters. The first question is what will you do if people you actually know, perhaps with young children in tow, who haven't eaten for days knock on your door? Let's make it even more of a dilemma. What if you have spent years of effort and a fair amount of your spare income building up your emergency supplies, and you know for a fact they made no preparations and don't even own a flashlight! So, what will you do?

If you turn them away and they know you have lots of stored food and emergency supplies inside, they just might come back later, only this time with other hungry neighbors more than willing to balance the scales. No doubt they will feel it's just not fair that you have so much, and they have so little—never mind they did absolutely nothing to prepare and believe it's your job to take care of them. Sound familiar? On the other hand, if you bring them all inside with open arms and a heart of charity, you may just end up running a boarding house for months with everyone wanting to know when you will have dinner ready!

What do the people adrift in a vast ocean on a leaking life raft do when they come upon more of their shipmates hanging onto wreckage and wanting to climb aboard, which will surely sink everyone. The answer will probably be different for each of us, especially since each circumstance will be different. Anyone preparing for a major disaster should anticipate they could face this exact situation and should set aside some supplies that will help others without risking their own family. While not the most tasty, dried beans and rice only need boiling water to prepare, are cheap, and have a long shelf life. Assuming the line at your door does not stretch around the block, this minimum food charity will at least avoid having to drive off your neighbors at gunpoint.

Security is also a concern, as history has made it abundantly clear, in every city, state, or country where government corruption or natural disasters have caused widespread food shortages and power outages lasting months, crime soon follows. A law-abiding person will turn to looting if their children haven't eaten for a week. It never takes long for established gangs and individual lowlifes to take out the proverbial pitch forks and torches and leave dark abandoned cities for the countryside seeking scarce food and supplies.

Barter will be the coin of the day and farmers' markets and swap meets always find a way to spring up whenever needed. I know one small group of individual families that live on a rural road that have already talked about coming together in the event of a major disaster or civil unrest. Several have small farms that can provide fresh vegetables, eggs, and meat, plus tractors to barricade the main access road with boulders. Several others are former military, one is a dentist, one a nurse, an auto mechanic, and several others are skilled tradesmen. It's interesting to note that no college professors or politicians were asked to join!

While you may not be lucky enough to live among this perfect combination of self-reliant individuals willing to help each other, you should at least start making the effort to discover exactly who you do live near. In today's fast-paced world, many young kids sit in their rooms glued to a computer screen and are not out playing, riding bikes, and making friends with the neighbor's kids as we did growing up. People living in the city today can go for years never knowing, let alone actually speaking to, their neighbors. Perhaps it's past due time to make some new friends?

Having nearby neighbors who are willing to come together and watch out for each other will go a long way in keeping criminal activity away, especially if some of your neighbors are prior military and have been collecting guns and ammunition while you have been collecting freeze dried food! No matter how well you have prepared for a future disaster, we all need to sleep sometime, and you absolutely must find a way to work with others until the power is restored and things get back to normal.

Another issue you will face is family. Have you invited distant family members to come and stay with you for a long holiday? Not only were you very glad to see them arrive, you were also very glad to see them finally leave, especially if they were there a week and brought young kids who spent all day on your internet account, while expecting you to feed them and clean up their daily messes.

Now imagine they are at your door after somehow managing to drive four hundred miles to reach you in a car now running on fumes, since all gas stations were closed, and the power has been out for days. Now imagine their kids are unable to power all their electronic games and cellphones, and there is no internet to keep them occupied from morning until night. You realize they could be with you for months and they have brought absolutely no survival skills or food with them.

What about sanitation and bathing for all these people for the next few weeks or even months? Perhaps it's time to give this some serious thought, especially if they are your brother, sister, uncle, or aunt, and might not share either your political or religious beliefs. To really top it off, they start blaming everyone else but themselves for the crises they find themselves in, which were the same exact things you have been warning them about for years while they continued to totally ignore your sage advice.

I'm not a psychiatrist, and we engineers are not usually known for our patience when dealing with others, so perhaps my advice should be taken with a grain of salt. But maybe there is a middle ground here. For example, regarding the general public and your local area, you need to keep a low profile, especially if you have been making lots of emergency preparations and have a large pantry. If you are the only house for miles around with every light on after a major storm thanks to your prior investment in a whole-house generator or solar backup system, then expect company!

I have a friend who is a serious prepper, and he and his wife have been working together for years to build their perfect retirement homestead. Their place includes multiple freezers full of meat and a backup well pump all powered by a large solar system. They have two huge walk-in pantries full of freeze-dried foods, bulk containers of wheat and rice, and shelves full of their canned vegetables and fruits from their large garden and orchard. Their homestead also includes a small fresh fishpond, lots of chickens, and a few turkeys. Everything they would need to live if society collapsed. A year ago they decided to reach out to their neighbors, since most houses on their long gravel road were spread far apart and they really didn't know anyone. They invited everyone living on their road to an open house weekend, and their hope was to display their prepping efforts in hopes they could convince their neighbors to also become more self-sufficient.

The event was well attended, and when my friend went into the kitchen to bring out more snacks, he could clearly hear several of the men talking out in the hall. Being unfamiliar with the home, the men apparently did not know he was just around the corner and heard them laughing and asking why should they prepare, they could just come there and take whatever they needed if things got bad since he obviously had enough for all. He learned a very painful lesson that weekend, as instead of hopefully making new friends, he just painted a target on his back.

You may not realize it, but these days there are lots of people around you who are serious preppers. However, they have learned the hard way how most people will not spend the time, effort, and money to become more self-sufficient. They now keep their prepping efforts quiet, as they do not want to become the local soup kitchen for the neighborhood when disaster strikes. The other common denominator is most preppers I know are very family oriented, generous with their time, and more than willing to help others.

Unfortunately, in today's world where news and television programs paint the prepper community as crazy morons standing guard at their log cabin door, barefoot, while holding a shotgun, it's no wonder they do not disclose their secret life to anyone. Why bother? Everyone knows that's what FEMA and the National Guard are for. They will rush in immediately to supply everyone with food, water, and blankets if disaster strikes . . . won't they?

Regarding close neighbors and extended family, you need to be firm in any conversations if they are already aware of your own efforts to be better prepared. Make it clear you do not intend for everyone to head to your house if disaster strikes. If the subject comes up, you need to turn it around and ask what they are doing to prepare for the unexpected. You can share your knowledge but take it slow. After all, it probably took you years to learn what you know now and years more to reach the level of preparedness you now have. Your first attempts to break the ice could simply describe how you keep some cash in each car in case the credit card system is down, and you are traveling and need gas or food. They may have just assumed the credit card systems would never fail!

If you are talking with a brother-in-law who has absolutely no clue, your first attempt to discuss the basics should not begin with a description of how to gut a deer! It's very hard to bring a total novice into the fold since most preppers only talk shop with each other. We have learned the hard way how the public will think we have three heads if we even mention the concept of being self-reliant. They are absolutely convinced the government will take care of everyone if disaster strikes, so why worry? While this may be partially true, more

SEE WHO IS AT THE DOOR

than likely if it comes to this, they may find themselves on a cot surrounded by thousands of other refugees inside a FEMA camp surrounded by a twelve-foot high steel fence and security cameras!

Assuming you have decided how best to handle the initial question as to who will be in your lifeboat, these next chapters will help you to at least make it as comfortable as possible.

CHAPTER 15

So, What Can We Do?

An EMP attack or solar storm can damage far more equipment and electrical systems than those addressed in these first few chapters. In fact, we may not know the cascading effect the loss of one system or device could have on other totally unrelated systems or devices until it actually happens.

Yes, the first half of this book does paint a pretty grim picture of what an EMP attack or solar storm can do to our nation, and our fairly civilized way of life. If our government and military cannot find a way to come together with the utility industry and reduce the nation's EMP vulnerability, and soon, then the devastation I have discussed may be inevitable.

Our country has many adversaries that would love nothing better than to destroy us if they thought they could get away with it and not suffer a major counterattack. Recent history has shown there are millions and millions of people in this world, including many living in this country who think the United States is just too rich, too powerful, and unfairly has too many natural resources. When much of the world is still in poverty or chains, adversaries can't wait to see the United States brought to our knees, and their speeches and internet postings display their hatred for us every day. Sadly, this includes some of our own elected state and federal representatives and many more in Hollywood and on university campuses.

This hatred caused by ignorance or corrupt leaders fanning the flames, almost guarantees an endless supply of current and future terrorists. These terrorists wear no uniforms and operate in small

SO, WHAT CAN WE DO?

groups or cells. They can inflect heavy damage to our nation's infrastructure without exploding a nuclear bomb. While the news media continues to cover up these events, the truth is there have already been multiple recent terrorists' attacks on our electric grid.

Substations have been completely destroyed using only a rifle, high-voltage transmission towers have been taken down by just cutting a few foundation bolts, huge fuel storage tanks at power plants have been damaged using crude homemade incendiary devices, and central grid control centers have been damaged by computer hackers operating from other countries.[1]

So far, these attacks have only caused regional grid outages, but due to their simplicity and difficulty to prevent, if multiple terrorists were able to coordinate these same attacks at multiple locations at the same time, it's possible to shutdown grid power to half the country, which could take months to completely restore. This is especially true if multiple extra-high voltage transformers are attacked since there are no spares and replacements can take up to a year to custom manufacture, ship, and install. It really doesn't matter what takes down our nation's electric grid, the results will be the same.

An electromagnetic pulse is not radioactive, although it is normally produced by the detonation of a nuclear device. There are nonnuclear EMP weapons, but they typically only damage the electrical systems in a limited area, not an entire state or country. There is an old joke where a city slicker asks a rural farmer where he would go if there was a nuclear attack. The farmer thought for a minute then said, "makes no difference, as long as you can say – what was that?" Unfortunately, with an EMP, the damage will be up to thousands of miles away and nobody will hear a thing.

Even though an EMP attack will most likely be caused by a nuclear bomb detonated at a high elevation above the United States, there will not be any nuclear radiation or explosive destruction at

SO, WHAT CAN WE DO?

ground level. In fact, at first there may be only a minor disruption to anyone not living in a city except for the loss of power, which all of us have experienced at one time or another.

The real impact will come later when store shelves are empty, and all water and sewer services stop. Once it is clear the power outage is widespread and could last months and not just a few days, things will take a dramatic downturn. So, the obvious question is what can we do? It is my hope the rest of these chapters will provide real common-sense solutions we can all use now, and I have tried to keep these suggestions as basic and inexpensive as possible.

Before continuing however, we should review what our government has been doing in recent years and oh yes, they are preparing, but they are preparing for their survival, not yours! I am hoping this will awaken those of you who still believe this government will rush in to save you. I'm sorry to report this isn't going to happen. You may be surprised to learn this country has had a master survival plan in case of nuclear war since President Eisenhower in the 1950s. Obviously, the emergency preparations have been significantly modified and expanded since then, and today it's called the Continuity of Government (COG) planning.

These disaster preparations started with the development of three primary underground facilities which include the recently upgraded NORAD Air Defense Command Center carved out two thousand feet below the surface of Cheyenne Mountain, in El Paso County Colorado. This underground bunker encompasses over five acres of floor space, with fifteen separate three-story buildings.

Each building sits on hundreds of heavy support springs and connected by flexible cabling to all utilities and emergency power systems. Blast doors at the surface help ensure the facility can survive a direct 30-megaton nuclear explosion, and the facility is completely shielded against EMP damage to its computers, communications, and emergency power systems. Having huge storage tanks of both fuel

SO, WHAT CAN WE DO?

and water, the facility was designed to house eight hundred people indefinitely. Obviously, this is far more people than it takes to monitor early warning radars looking for Russian missiles. So, who else has a free ticket to get in when hostilities begin?

The two closest underground government shelters near Washington, DC, are the Raven Rock Facility located near Liberty Township in Pennsylvania on the border of Maryland and near Camp David; and the Mount Weather facility located near Berryville, Virginia. The Mount Weather facility replaced the older and much smaller nuclear bomb shelter built further south under the Greenbrier Resort located in White Sulphur Springs, West Virginia. Mount Weather serves as the main emergency operations center for the Federal Emergency Management Agency (FEMA). It's also intended to house the highest level of our government in the event the president and his cabinet must be relocated out of Washington, DC.[2]

Nine federal departments maintain a duplicate backup operation at Mount Weather, and each has the backup data and communications to take over immediately if their primary operations in Washington, DC, are threatened by nuclear war or civil unrest. The underground data center located at Mount Weather contains backup copies of the personal records of almost everyone in the United States. These files also contain a list of those citizens recognized as "vital" for restarting our country after civil unrest or nuclear exchange, plus a special list of those citizens considered to be a "potential risk" to any declaration of martial law during a civil crisis that could include an EMP attack. Both lists are constantly being updated.

This underground complex is actually a small city with paved roadways, sidewalks, and a battery-powered subway system linking twenty separate multistory office buildings, multiple cafeterias, food storage, a hospital, dormitories, a power plant, and one of the world's

SO, WHAT CAN WE DO?

largest and fastest mainframe computers.³ The underground facility also includes 500,000 gallons of fresh water, a 90,000 gallon per day sewage treatment plant, and can sleep 2,000 people. During non-emergencies, Mount Weather has a daily workforce of approximately 1,000 employees, most stationed in the part of the facility that is above ground.

Federal documents and congressional committees refer to this FEMA Headquarters only as the "special facility," but "Mount Weather" is the term most often used as it originally was a weather station located on top of this Virginia mountain. The primary emergency access to Mount Weather is by helicopter and underground tunnels, plus a small private airstrip located down the mountain in a valley a few miles away.

Mount Weather is the operational center of FEMA, but there are now over one hundred additional Federal Relocation Centers it directs which are scattered across Pennsylvania, West Virginia, Virginia, Maryland, and North Carolina. There are also multiple smaller underground facilities located in many other states that are part of this country's Continuity of Government (COG) program and their secure military communication systems.

Each facility is designed with a bug-out plan for each specific branch of our government, with Mount Weather housing the President and his Cabinet, and Ravin Rock housing the various branches of our military. Raven Rock serves as the emergency operations center for our military and is unofficially known as the Underground Pentagon. Its rock tunnels house multiple three-story spring supported buildings like the Cheyenne Mountain facility. It also has enough stored fuel, water, and food to accommodate up to a thousand high-ranking military leaders.

Another multi-million-dollar COG facility having both above and below ground infrastructure sits on top of Peters Mountain a few

SO, WHAT CAN WE DO?

miles northeast of Charlottesville, Virginia. It is one of the many surviving Automatic Voice Network (AUTOVON) communication relay facilities originally built by AT&T in the 1960s to provide a totally separate and hardened military phone network that does not utilize any of the United States civilian communications lines. Other than a few small metal storage sheds, a parking lot, and a helipad above ground and surrounded by dense woods and very secure fencing, the bulk of this communications facility is thirty feet below ground and protected by EMP copper shielding.[4]

The old Allegany Ballistic Laboratory (ABL) located in Ridgeley, West Virginia, a few miles east of Washington, DC, has undergone multiple recent renovations and is now a huge campus on the top of a very isolated mountain and houses FEMA's main data center plus multiple separate support facilities.[5]

The extensive underground parts of the ABL campus houses backup facilities for multiple government security agencies including the National Archives. This agency compiles and prints the daily Federal Register, and no law, regulation, or new standards get implemented until printed in this daily publication. While normally printed by the Government Printing Office in downtown DC, this backup facility located in the remote ABL facility maintains a mirror image backup to everything recorded and printed at the downtown DC location.[6]

While each of the twelve Federal Reserve banks maintain their own backup underground facilities, until 1997 the Richmond Federal Reserve's backup location was the Mount Pony facility located just east of Culpepper, Virginia. The main above ground administration building looks like a semicircular hotel built on the side of a wooded hillside and is easily visible from Route 522. However, behind all those glass windows are motorized shutters of five-inch thick metal shields and 140,000 square feet of below

SO, WHAT CAN WE DO?

ground EMP proof facilities. Its empty offices did stand ready with duplicate desks and constantly updated data files to support sixty relocated staff to oversee all daily electronic funds transfers from all Federal Reserve Banks.

The underground facility has backup generators and until 1997 stored food and water to house five hundred additional staff and one of the world's largest vaults originally holding $1 billion in cash to reboot the economy during an estimated two-year recovery period.[7] This underground vault and four-hundred-foot long underground bunker are now home for all of the Library of Congress's rare film archives – at least that's what we are told!

Every government agency below this main group also have been mandated over the years to develop their own backup facility that mirrors their day-to-day operations. While these are separate facilities, each have their own emergency preparedness plan and are located within an easy helicopter ride from Washington, DC.

Special identification cards are issued to each official in each government agency deemed to be "critical" in the COG plan, which instructs any police they encounter when rushing to their assigned underground facility to provide immediate assistance. These major government and military officials are also assigned a helicopter pilot, instructed to stop what they are doing and head to a designated pickup location if they are paged. There are also extensive underground tunnels leading to other secure facilities for use in evacuating officials out of Washington, DC. I don't know about you, but I checked my wallet today and did not find one of those "get out of jail free" cards!

FEMA has been constructing a large number of above ground detention facilities that are currently standing empty, but ready to house thousands of civilian detainees. Originally called the Readiness Exercise 1984 Plan, or REX-84, it was thought these facilities would

SO, WHAT CAN WE DO?

be needed in case there was a mass exodus from Mexico into the southwestern United States.[8]

The official number of these detention centers is unknown since FEMA will not confirm, but unofficial estimates believe these detention centers number in the hundreds. It is known that many were built by renovating vacant shopping centers or abandoned malls. Most renovations started with blocking all the windows and installing high security fencing and camera towers around their large paved parking lots.

We know these do exist as several of these FEMA camps were recently used to temporally house several thousand underage migrant children that crossed our southern border without parents. Unfortunately, it is feared these currently still empty detention centers may soon serve a less humanitarian purpose.

Many believe that there will be a roundup by FEMA working jointly with our military or UN troops to house thousands of individuals currently on a "watch list" of those citizens the government does not want unaccounted for and roaming free if martial law is ever declared. It is thought this constantly updating list will include those citizens with a history of anti-government activities or internet postings, anyone that has had prior military training, and those known to keep weapons within their home. Keep in mind Homeland Security legislation allows rounding up anyone suspected of terrorism activities without a trial, or even notification of relatives regarding where they are being held or why.

Actually, this is not the first time this country has used detention centers. Immediately after the Pearl Harbor attack in 1941, over twelve hundred community leaders of Japanese ancestry were arrested and their bank accounts frozen. Curfews were put in place for all cities along the Pacific coastline of California and anyone of Japanese descent caught out after dark was arrested. A year later the

SO, WHAT CAN WE DO?

War Relocation Authority (WRA) was enacted and all Japanese Americans were ordered to report with all family members to multiple relocation centers and their cars were confiscated. Over 120,000 Japanese Americans spent the next three-years detained in ten different detention centers located in western states. At the end of the war most found their homes, businesses, and most belongings left behind were missing or sold, never to be recovered.

There had been no trial and no reading of their rights. They were told to just leave everything behind, and you have two weeks to show up and be fenced in for the next three years. If there is a major grid down event caused by a nuclear EMP, there will be major confusion between our elected leaders, the military, and a news media promoting every possible retaliatory action, no matter how ludicrous. Couple all this confusion and hysteria with weak leaders and we could easily see government agencies anxiously waiting on instructions to round up whatever group of American citizens will be deemed a threat this time.

It is not known how FEMA plans to use these new detention centers in the future, but many suspect they could be used to round up "undesirables" during a future declaration of martial law. It is expected at least some centers will be used to house large numbers of citizens wandering the streets with no place to go and no food or water after a major grid down event. I hope it is clear to you by now that the "so what can we do?" question can be answered by being as prepared as possible for a future power outage and resulting civil unrest that may last months, if not a year or more.

So, what does all this mean? It means our government has realized since the 1950s when this COG planning first began that it is just not possible for any central government agency to take care of millions and millions of people needing food, water, and shelter after an EMP attack, so the only thing they could do is make sure the

SO, WHAT CAN WE DO?

government survives and then reboots after 90 percent of the population has died off.

The COG planning is to ensure government and military officials will survive for months in underground shelters with plenty of food and water, and heavy security, after a nuclear attack, civil unrest, or breakdown of our nation's infrastructure. Those wondering the streets searching for discarded food will eventually be rounded up and bussed to one of these FEMA camps, assuming they are not attacked first by the roving gangs stealing everything of value and killing anyone who has the misfortune of crossing their path.

As this final draft was on its way to the publishers, the Coronavirus, or COVID-19 pandemic had just reached the United States. While I expect the virus to have run its course by the time this book reaches bookstores and life hopefully has returned to at least a modified version of "normal," the government did implement for the first time since 9/11 the Continuity of Government emergency plans in mid-March 2020. The Continuity of Government Readiness Conditions (COGCON) include four levels of readiness, and at the time of this final editing the United States was moving from COGCON 4 to COGCON 3 (see the table on the next page).

In the event of an EMP attack and the resulting long-term power outage, if you live in a large city your immediate action should be to just get out. Of course, this also means you will need somewhere to go and a way to actually get there! Preparations for a major grid down event can take months of planning and assembling supplies. However, these steps are very easy to do and are not expensive. Obviously, you can't prepare for every possible threat. The goal is to find ways to prepare now to have an acceptable comfort level while all grid power and normal communications could be down for up to a year.

SO, WHAT CAN WE DO?

> **COGCON READINESS RATING**
>
> COGCON 4: Federal executive branch government employees at their normal work locations. Maintain alternate facility and conduct periodic continuity readiness exercises.
>
> COGCON 3: Federal agencies and departments Advance Relocation Teams "warm up" their alternate sites and capabilities, which include testing communications and IT systems. Ensure that alternate facilities are prepared to receive continuity staff. Track agency leaders and successors daily.
>
> COGCON 2: Deployment of 50-75% of Emergency Relocation Group continuity staff to alternate locations. Establish their ability to conduct operations and prepare to perform their organization's essential functions in the event of a catastrophic emergency.
>
> COGCON 1: Full deployment of designated leadership and continuity staffs to perform the organization's essential functions from alternate facilities either as a result of, or in preparation for, a catastrophic emergency.

This means you will need to store up to six months of easy-to-prepare foods for each member of the family; a large supply of stored water and a way to filter contaminated surface water; a simple way to prepare meals without a working kitchen; a simple way to maintain a level of personal hygiene and wash clothing; a simple way to deal with sanitation; a simple way to keep reasonably warm during cold winters and not over heated during hot summers; a simple way

SO, WHAT CAN WE DO?

to power low-voltage lighting, a radio, and pump water; and perhaps some limited refrigeration.

The following chapters will deal separately with each of these topics, with the exception of self-protection. There are a large number of books and articles by others on emergency medicine and self-defense which I strongly recommend adding to your emergency preparedness library. Several are listed in the Recommended Reading section located in the Appendix.

The electric grid can fail for a variety of reasons not related to a nuclear EMP attack or solar storm. Although an EMP will cause the most damage and require the longest time to recover, power outages have lasted weeks – and even months for those hardest hit – from hurricanes, floods, forest fires, equipment failure, and even operator error. My goal is helping you learn the benefits of using battery-powered devices after most grid-powered appliances have been damaged or have no 120-volt AC power to operate after an EMP attack. However, these same recommendations will help you prepare for any extended power outage regardless of the cause.

Not all vehicles will operate, gas stations could be closed, and most main roads could be blocked by major traffic grid lock and abandoned vehicles. Plan to have multiple alternative routes to reach your emergency location that avoids bottle necks and possible roadblocks typically located near major interstate exits, bridges, and other traffic bottlenecks around cities.

Once these basic needs have been satisfied, there are many other things that will make life easier and safer during a grid down event. This includes having a way to communicate with neighbors to increase security as there really is safety in numbers; a way to communicate with friends and family when both cell service and landlines are down; a way to stay informed of both regional and national news; and a way to receive emergency alerts when all

SO, WHAT CAN WE DO?

normal forms of radio, television, and internet systems have completely failed or out of service.

In addition to needing a way to power all of these alternate forms of communication, life would be far easier during a grid down event if there was a way to have lighting, limited refrigeration, clean water, and some entertainment long after the grid and your generator have failed.

The next few chapters will help you do all of the above, and the ways I will describe are both simple and fairly low cost. In addition, you do not need any special skills or training. All recommended electrical devices I suggest are battery powered and operate on safe low-voltage power. In addition, everything can be packed up and taken with you if you are forced to leave. It's time to take action, so let's get started.

CHAPTER 16

Transportation after an EMP Event

Preparing your vehicle for a possible EMP attack means more than just getting it started again. When traveling on vacation on a very hot July in our truck and RV we suddenly found ourselves in a ten-mile backup on the Interstate. Everything just stopped, and all we could see were cars and trucks for miles in front and miles behind in all lanes.

To say it was hot and humid is a major understatement, and after the first few hours in the hot sun people around us, especially older people, women, and kids were needing a bathroom break and had nothing to drink. The lack of water was especially taking its toll on the very young kids whose parents no doubt thought they would be on the road just long enough to go to the store. Since nobody was going anywhere, we opened up the RV for emergency bathroom breaks for those around us and took out the ice and bottled water for the kids so life became almost normal again. This proves having some bottled water at all times is an absolute necessity, even if you think you will only be gone fifteen minutes!

Chapter 11 reviewed how an EMP can affect cars and trucks, while this chapter will provide ways to restart an EMP-disabled vehicle. During actual testing of a wide range of vehicles in EMP test chambers, it was learned that most vehicles were not damaged by a moderate level of EMP energy *if* their engines were not running during the test. However, most engines shut down when exposed to

TRANSPORTATION AFTER AN EMP EVENT

levels of EMP energy that was significantly below the maximum levels expected from an actual nuclear explosion at high altitude. If you were driving when an EMP attack occurs, the engine will probably quit along with all the other vehicles around you. This means the immediate priority is trying to coast to the shoulder if possible and get out of the way. It's doubtful cell phone service will be working, so don't expect to call AAA or a relative for help.

This actual EMP testing also found many of the vehicles that failed were able to function again and the engine restarted after the vehicle battery was temporally disconnected, which basically rebooted all of the computer controls. Unfortunately, it's almost impossible for some drivers to even find the battery in their compact car, let alone actually have the tools to remove the access panel and disconnect the battery terminals.

You should keep a good selection of spare fuses in the vehicle, as the high voltages induced into the wiring by an EMP will cause fuses to blow which will stop all electrical systems. If you have the skills and tools on hand to disconnect the battery and replace failed fuses, it's possible your vehicle could be made drivable again. However, if initial attempts will not get you going and you are stranded, then perhaps it's time to change your shoes, grab the backpack, and head out!

Actual testing at moderate levels of EMP energy did cause some damage to almost every vehicle even if they could be driven again. In addition, most of this testing did not take the EMP energy up to the level expected during a real nuclear EMP event, so actual EMP vehicle damage is unknown and will most likely vary depending on where you live, vehicle orientation, and if parked in a garage or left out in the open.

There is a real chance you could be standing next to a disabled vehicle in the middle of nowhere after an EMP hits. This means you will be walking, and not like the walking you do each day going to lunch! We are talking about walking twenty, thirty, or more miles to reach a safe place, and this will not be easy even for those of you in good physical health.

You should keep bottled water and real hiking boots with padded hiking socks at work and in the trunk of each vehicle. Tennis shoes, flip-flops, and high heels will be totally useless. It's also possible it could be very hot, very cold, raining, or snowing when heading out on foot. Along with your backpack and hiking boots you should include protective clothing with each backpack that is switched between winter and summer each change of season.

Even if most vehicles can remain operational after an EMP attack, the real transportation crisis will be no fuel due to closed and dark gas stations, disabled traffic lights, and highways blocked by abandoned vehicles. While our truck guzzles fuel when hauling our RV, I also have a compact car that gets forty-eight miles per gallon. This means I can travel about five hundred miles on ten gallons of regular gas, and since it's not a hybrid or battery-powered, it has a much simpler electrical system.

While having a four-wheel-drive truck and RV make a great way to travel if forced to evacuate, having a smaller vehicle that gets incredible mileage could be extremely handy for short runs when gasoline could be rationed for many months. I keep multiple five-gallon gas cans full at all times and rotate these cans with the lawn tractor fuel cans during the summer to keep everything fresh. I add a fuel stabilizer to each gas can when refilling as engines can be very hard to start once gasoline is over six-months old, even with the fuel stabilizers.

One form of low-cost transportation we tend to forget about while using all the many forms of today's fast transportation systems is the lowly bicycle. High tech materials have resulted in a resurgence of these fairly basic, and totally EMP proof, ways to get around. Extremely strong and lightweight frames have significantly reduced

TRANSPORTATION AFTER AN EMP EVENT

overall weight, while fat knobby tires and large tractor style padded seats makes trail riding much more comfortable, especially for those of us who are older. Since these are a great low-impact form of exercise, this is another EMP prep you can use and enjoy now.

For more distant travel, a small gasoline-powered trail bike or even a battery-powered off-road electric bike could be a great way to run errands or check on the neighbors after any type of major disaster takes down all normal forms of vehicle transportation. These can easily maneuver around downed trees and storm damage that would stop all full-size vehicles. The electrical systems on these motorized bikes are very basic, and if any longer wires to headlights or accessories are disconnected from the motor housing when stored, they should be fairly EMP proof.

It's not expensive to keep a backpack, hiking shoes, and a few tools in each car. It's also important to keep both a detailed city street map and state map in each vehicle. The GPS system on which most of us now depend may not function after an EMP event, and you may need to find alternative routes to a destination due to blocked roadways, high water, or fast-moving fires.

If you anticipate needing to travel to a distant location after a grid down event, which could require driving on long stretches of unimproved backroads or mountain hiking trails, getting lost is a distinct possibility. There is a very good chance the GPS satellite system will be disabled, and you probably will not have a large collection of paper maps. But there is another solution.

I have a battery-powered handheld Garmin eTrex GPS Navigator. While it does rely on a GPS signal to accurately pinpoint your specific location, the manufacturer offers MicoSD cards with preloaded all-terrain maps. These stay in the device and the unit does not need an internet connection or GPS signal to track you travel progress or display these very detailed maps.

TRANSPORTATION AFTER AN EMP EVENT

In addition to displaying all major highways, the color screen displays backroads, hiking trails, small streams, fire trails, terrain elevation contours, and other helpful landmarks not included in a normal GPS unit, even with a working internet. The unit has other screen views including a compass and satellite images you can download ahead of time of where you plan to go in an emergency. These devices are EMP proof due to their small size and will operate weeks on one charge.

If you have the extra funds, having a backup off-road vehicle or perhaps a powered trail bike will always come in handy. Even if never actually needed for a future crisis, they are fun to drive, as long as you are not trying to outrun the zombies!

CHAPTER 17

Will Your Generator Survive an EMP?

I remind people in my seminars that my promoting the use of battery power does not mean I'm against owning a generator. In fact, we have a whole-house generator to back up our home's solar power system, an RV with built-in backup generator, and a portable generator, which is occasionally needed for remote construction projects. Generators do make life easier during a weeklong power outage after a hurricane. However, I do not expect any generator to still operate after an EMP event, or to have an endless supply of fuel if it does. Our whole-house 12-kW generator easily powers our 3-ton central AC unit and still keeps several refrigerators and all lights operating. A larger 20-kW whole-house propane generator can power a 5-ton central AC unit, or two smaller heat pumps, but will drain a five-hundred-gallon propane tank in a week if operated continuously. This means even if a backup generator survives an EMP attack, their usefulness is limited to only a week or two before depleting the fuel supply when refilling could take months.

All portable and fixed residential whole-house generators sold today contain an electronic ignition system and additional electronics to smooth out the fluctuating voltage, filter out electrical noise, and display engine diagnostics. The generator's electronic control board providing these functions contains the same microchip components as all other electronic systems that EMP testing has easily destroyed. While an all metal generator housing can partially attenuate the E1

WILL YOUR GENERATOR SURVIVE AN EMP?

pulse traveling through the atmosphere from reaching the sensitive electronics inside, the number, size, and location of any enclosure openings will reduce its effectiveness.

Fig. 17-1. A whole-house generator only works as long as you have fuel.

If you have some wiring or ham radio experience, adding properly sized metal oxide varistors (MOV) as discussed in chapter 32 and high-frequency blocking ferrites as discussed in chapter 33 will help reduce EMP damage to a generator. However, an easier solution is just to keep all circuit breakers between the generator, the home's electric panel, and especially the connection to the electric grid turned off when not in use. Of course, this will not let a backup generator automatically take over house power during a normal power outage without manual intervention. Most likely the generator's electronics will be seriously damaged by either E1 traveling through the air, or E2 and E3 coming in on the power lines. Keeping a spare generator control board in an EMP proof container as described in chapter 36 might be good backup insurance, but even if you are able to make the repairs and switch out the control board, there is still the issue of fuel. Propane deliveries may not resume for months and gas stations could be empty and closed for long after a grid down event.

WILL YOUR GENERATOR SURVIVE AN EMP?

Multiple Russian nuclear bomb tests over land in 1962 imposed high-voltage surges on long utility lines which fed back into the coil windings of several large industrial generators connected to these transmission lines. This resulted in significant damage in the coil windings of these generators as the arcing between the windings caused a breakdown of the dielectric properties of the wire insulation. Months after these nuclear bomb tests were completed, several generators appeared to operate normally when tested, but when the generators were actually put back in service, most generators failed.[1]

While completing a recent magazine article on portable generators, I tested the very small and inexpensive 900-watt PowerSmart generator exclusively sold by Home Depot, and the almost identical 900-watt GATOR generator sold exclusively by Harbor Freight. Both models have a two-cycle engine, so you need to mix oil with the gas. Also there is no crankcase full of lubricating oil, so nothing to leak. These can be stored almost anywhere and in any position. These small hand-crank generators can be more than a little temperamental to start, and their limited wattage output and poor voltage regulation is not suitable to power large appliances or sensitive electronic equipment, but what do you want for $110.00?

After a grid down event, these could be brought out of storage, filled with mixed fuel, then connected to multiple outlet extension cords to recharge every cell phone, iPad, portable radio, and walkie-talkie in the neighborhood! These also include a separate 12-volt DC charging outlet and the one-gallon gas tank provides a five-hour run time on a single fill up. A generator this size is perfect for recharging multiple small electronic devices or power tools at the same time. Being EMP proof, having low fuel consumption, and basically cheap, it's worth owning one even if you already have a larger whole-house generator. In fact, at this price I would buy two and have a perfect replacement for every single part!

WILL YOUR GENERATOR SURVIVE AN EMP?

Having a working generator to carry you through a storm related power outage lasting a few days is very convenient. However, when preparing for a power outage that lasts months, regardless of the cause, a better long-term solution is to significantly reduce your dependence on electricity and have battery-powered devices to bring out of the closet and take over when grid and generator power are lost. The next few chapters will discuss multiple ways portable battery-powered devices can provide a comfortable off grid lifestyle, and there are many ways to keep these charged during a power outage lasting up to a year.

CHAPTER 18

Lighting after an EMP

Without grid power, you will still need a way to have light in any room you occupy on a regular basis, but this may not need to be the higher light levels required during normal times. Areas for meal preparation usually require the highest light levels. Switching ceiling and task light fixtures to lower wattage LED bulbs now will significantly extend the number of days you can operate from any backup power system. You can find an LED bulb for almost every size and style light fixture imaginable, including 12-volt DC powered fixtures designed for RVs and boats. Although more expensive, replacing all 60-watt or larger incandescent light bulbs with LED lamps means your backup power system just got downsized by a factor of ten, or will provide the same light levels ten times longer. This will also reduce your current monthly electric bill while utility power is still available.

When selecting light fixtures for use during a grid down event or to minimize electrical usage during normal grid-powered times, avoid light fixtures having a frosted lens and non-reflective interior. Look for fixtures having a clear glass lens or bulb enclosure, and polished metal or a mirror type interior to maximize the light being projected from the fixture. This will allow using lower wattage bulbs and is critical when using LED type lamps.

Solar-powered LED walk lights can be found in almost every yard and driveway these days, yet nobody realizes how these make perfect emergency lighting systems during a power outage. They are charging outside in the sun all day, then they provide light throughout the night. LED walk lights are rated in lumens where one lumen is the equivalent light from a single candle as measured at a distance of one foot.

Solar walk lights having a 25-lumen or higher rating are ideal for a kitchen, dining room, and family room. Smaller solar walk-lights in the 8-lumen range should provide adequate light levels in halls and stairwells. Bathrooms will require 12-lumen or higher rating for adequate lighting at sinks and mirrors. Make sure the models you select have clear lenses and polished-mirror reflectors to maximize the light distribution. Due to small size and no external power wires, these LED yard lights are totally EMP proof.

I have noticed some of the newer solar yard light designs seem more concerned with their daytime architectural appearance than their nighttime light distribution. Those with multiple wide arms around the perimeter of the lens to support the solar cell top produce major shadows in the illuminated area at night. Not only do I find this distracting during normal nighttime sidewalk illumination, this is totally unacceptable for emergency interior lighting. Be sure to select only those designs with minimum light blockage around the perimeter of the clear lens. Since you will be buying these for the additional purpose of providing emergency backup lighting, avoid models with a lower-capacity battery and flimsy construction.

Although designed to be installed in the ground with a spiked tube support, a more practical installation involves driving a 3/4-inch PVC conduit in the ground with two feet remaining exposed. Be sure each location provides the normal exterior illumination you need, while being free of shade during the day. Remove and discard the spike end and insert the tube support of each solar light into the conduit. This makes it easy to remove each evening during a power outage and take inside.[1]

LIGHTING AFTER AN EMP

Fig. 18-1. Solar walk lights make great emergency indoor table lamps.

Using discarded wood blocks, mount a short piece of the same size PVC conduit used for the outside support into the center of each wood block creating a lamp base. You will need one for each room requiring emergency lighting. Each morning simply remove the solar lights from their wood bases and return to the yard for recharging. This is an ideal power outage task for the kids. Even if you still have fuel for your generator, there is no need to waste it when this simple do-it-yourself project can provide all the interior lighting you need during any power outage, regardless of how long it lasts. The rechargeable batteries that come with the higher-cost solar walk lights typically have a two-year life under normal usage. Try to standardize on models requiring the same size batteries and be sure to have spares.

LIGHTING AFTER AN EMP

Every home should have multiple LED flashlights that are fully charged and located for easy access in the dark. LED flashlights that are powered by rechargeable batteries are a must have. In chapter 28 I address the benefits of today's longer-lasting rechargeable batteries. During extended power outages, generators and emergency backup systems will fail or need to limit run time each day to save fuel. Without grid power, your home will be extremely dark, and this can occur unexpectedly. Having an LED flashlight beside every bed and main entry door is critical for your emergency power outage preparations.

You will also want an LED flashlight in each vehicle and bug-out bag. I have found that most of the lower-cost LED flashlights require three of the smaller AAA size batteries, while the slightly larger and more expensive LED flashlights will use two AA size batteries. Throughout this book I stress the importance of using rechargeable batteries and making sure all of the battery-powered devices you own require a limited number of battery sizes. I would avoid entirely any device that requires the small AAA size batteries as these are just too small to power anything for any length of time. Probably the most useful battery size that fits almost all digital cameras, flashlights, electronic games, and portable radios is the slightly larger AA size battery.

Originally sold for the camping and RV enthusiast, my favorite room light during an extended power outage is an LED lantern, similar in design to the old propane or kerosene lanterns, but smaller. Unlike a single-direction flashlight, the LED lantern is designed to project its light 360 degrees, in all directions, and most models have a much larger battery capacity than a typical flashlight. Some models include a fold-up hook for hanging and can be easily carried from room-to-room and hung from a ceiling hook.

Fig. 18-2. Battery-powered LED lanterns easily illuminate a room and some models include a solar charger.

When suspended near the location of any existing ceiling light fixture, a single LED lantern will provide acceptable lighting levels throughout most bedrooms, family rooms, and bathrooms. Adding a small hook near each ceiling fixture to hang these lanterns during a power outage is an easy way to prepare now, and these hooks will most likely never be noticed. When selecting an LED lantern, make sure it uses the same size rechargeable batteries as your other battery-powered devices, with C size batteries being the most common battery used in the larger LED lanterns. If the batteries are designed to stay in the lantern and be recharged from a 120-volt AC power supply, you will need to order the optional 12-volt DC adaptor so they can be charged from multiple 12-volt DC power sources during a grid down event.

LIGHTING AFTER AN EMP

Try to purchase the same model of LED lantern to use in multiple rooms in addition to using the same size batteries. At a minimum you will probably want a hook-suspended LED lantern in the family room, kitchen, and each bedroom. Smaller models or solar walk lights can be used in corridors, bathrooms, and stairs. These LED lanterns and LED flashlights are not affected by an EMP or solar storm as long as not connected to a battery charger when an EMP hits.

I have tested multiple styles of LED lanterns and LED flashlights in real EMP chambers and due to their small size, they were not damaged unless they were connected to a charger or long charging cable. However, I also tested several types of LED string lighting, which have multiple LED lamps along the support strip, and these did suffer damage. If you look closely at the strip, there are multiple tiny microchips embedded along the entire length between every few LED lamps which regulate the voltage and current going to each LED lamp. These microchips definitely are

Fig. 18-3. LED strip light before EMP test. Note all lamps are illuminated.

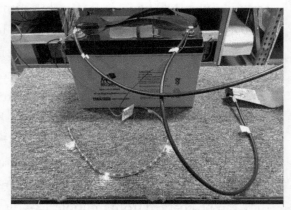

Fig. 18-4. LED strip light after EMP test. Note multiple lamps no longer illuminated.

LIGHTING AFTER AN EMP

damaged by an EMP due to the wiring strip being long enough to collect EMP energy.

During an extended power outage, regardless of the cause, having at least some light in each room can make a difficult time at least bearable. Do not think that owning one or two flashlights will do the job either. Your first goal in being prepared after taking care of food and water should be multiple battery-powered LED lights and a way to keep these charged.

If you need to maneuver outside at night to check on pets, bring in firewood, or make vehicle repairs, I have found the small LED headlamps to work great. The strap easily fits around your head or a cap, and these small lights are incredibly bright. Some models use the larger AA batteries I recommend, and some have rechargeable batteries and a USB charging port.

The main thing to take from this chapter is you need multiple LED flashlights and lanterns, and a way to keep them operating no matter how long the grid is down.

CHAPTER 19

Powering Communications after an EMP

Having a portable battery-powered all-band radio can be a real lifesaver after an EMP attack and these do not require the grid or a generator to work. If you decide to purchase a handheld radio, select a model having both a built-in speaker and separate ear buds. While the speaker allows others to listen at the same time, switching to the lower-power ear buds when alone provides much longer battery life. While FM transmission provides clear stereo broadcasts that are virtually free of static, FM transmission is limited to line-of-sight distances. Depending on the height of the FM station's antenna, the reception area rarely exceeds 20 or 30-miles, especially in mountainous terrain.

With the current popularity of talk radio, manufactures of battery-powered AM radios are starting to include higher-quality AM antennas and more sensitive AM receiver circuits to maximize reception of distant AM stations. In addition, some battery-powered radios will also receive several shortwave radio bands, allowing the reception of overseas programs and news broadcasts typically available in English as well as their country's language. Make sure the battery-powered radio you purchase is advertised as having improved distant AM reception and also receives multiple shortwave bands.

While we normally think of a battery-powered radio as something you can carry in the palm of your hand, several tabletop models are available with high-capacity rechargeable batteries and

POWERING COMMUNICATIONS AFTER AN EMP

large speakers for room filling sound. Most models will also work on grid power when available to save battery charge. Like the families of the 1930s gathered around their radio in a large wooden case the size of a desk, a battery-powered radio may be your family's main source of news and entertainment when all other forms of electronic communications are down.

Fig. 19-1. Tabletop radios are available to operate on both grid power and internal batteries.

Look for a portable radio that requires either the AA or larger C size batteries. Avoid any radio requiring the tiny AAA or large D size flashlight batteries as it is doubtful these sizes will be needed for any of your other battery-powered devices. Always remember, during a real grid down event, keeping your battery-powered devices operating and fully charged will most likely require playing "musical chairs" with batteries, adaptor cables, and various charging devices, so keep it simple and standardize wherever possible.

Cell phone companies have increased their on-site generator fuel storage based on recent experiences with longer storm-caused power outages, but sooner or later they will run out of fuel and fuel delivery will be problematic during an extended power outage. After an EMP event, testing has shown there will be at least some damage to the electronic components in each cell tower. While many can be put

POWERING COMMUNICATIONS AFTER AN EMP

back in operation by a simple board replacement, there still may not be a functioning grid to keep them running.

Keeping your cell phone fully charged is extremely important as most cell phones now include the ability to send and receive text messages, which usually can get through during limited reception conditions when normal voice calls will not. A quality cell phone may offer other helpful forms of communication during an emergency including accessing the internet for web-based news, sending and receiving e-mail to distant relatives, and photo documentation of events.

While all cell phones come with a standard 120-volt AC charger, it's critical that you purchase several optional 12-volt DC car chargers and place one in each vehicle and in your bug-out bag. While the cellular service will eventually end during a grid down event, it should still be available in most regions of the United States for the first few weeks using generator power unless the outage was caused by an EMP attack, then all bets are off.

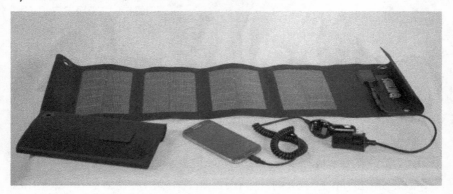

Fig. 19-2. All cell phones can be recharged using a foldup solar charger.

One of the easiest ways to keep any cell phone or personal computing device fully charged, regardless of how long the grid is down, will be a foldup solar charger. When folded up for storage and not connected, these will normally survive an EMP event. In

POWERING COMMUNICATIONS AFTER AN EMP

addition, they are fairly inexpensive and come in many different sizes. While the smallest 5-watt solar charger will eventually charge up a cell phone, a 10-watt or larger model will cut charge time in half. Smaller wattage solar chargers typically have a 5-volt USB type charging outlet, while larger wattage solar chargers capable of charging a laptop computer have a 12-volt DC charging outlet.[1] There are also better-quality solar chargers that provide both a 5-volt USB and a 12-volt vehicle style outlet. I keep a 10-watt foldup solar charger in each of my bug-out bags and one stored in the glove box of each vehicle.

The Family Radio Service (FRS) channels are in the ultra-high frequency range of 462 to 467 MHz and utilize low-noise FM transmission which has less interference than AM radios. Unfortunately, a FRS radio does not have the range to let you talk to somebody in the next city. Even with limited transmit range, these are a great way to set up a local neighborhood watch after a grid down event or period of civil unrest, and they are very inexpensive.

Although first set up in 1996 for private family and farm use before the wide use of cell phones, businesses found handheld radios to be a perfect way to communicate with employees and service technicians working in large warehouses and on multi-building campuses. These walkie-talkies have multiple channels, very clear reception, and they operate similar to a cell phone when using their privacy paging tones.

Even though all cell phone and land line phone service may be lost, you still need to be able to communicate with others. While limited in range due to the 1/2-watt transmit power, a quality FRS walkie-talkie will still let you talk to nearby neighbors and there is safety in numbers, especially when there is a risk of roving gangs. While most FRS radios are advertised to have a range in miles, this is only true over a flat open field. Do not expect to talk with anyone further away than a city block, especially if your area has lots of hills

POWERING COMMUNICATIONS AFTER AN EMP

or wooded terrain. The lower-power FRS radios are typically sold in sets of four or six, which avoids the problem of your neighbors using their own radios that may not have the same channels or paging codes as you are using. Being able to hand out battery-powered walkie-talkies with matching preprogrammed channels to neighbors during a grid down event creates a great local early warning system.

The demand for longer-range commercial communication has resulted in a more powerful line of extremely rugged hand-held radios. The General Mobile Radio Service (GMRS) walkie-talkies operate on the same first seven channels of the fourteen channel FRS radio band, plus an additional eight channels immediately above the FRS channels.

These more expensive GMRS radios have up to 5 watts of transmit power and do require a station license. However, this license does not require passing a test as it's only a registration process. I have a pair of these small handheld GMRS radios which I now use when in remote areas that still do not have any cell service. I have found these to have a true range of two-miles in a very mountainous and wooded area, and they make no sound unless you are being paged. While the smaller wattage FRS radios are cheap and suitable to talk with a nearby neighbor, a pair of GMRS walkie-talkies will be invaluable for greater distances and in more wooded terrains. Both the family FRS walkie-talkies and the commercial GMRS radios are EMP proof due to the very short antennas and no external wires unless sitting in a charger.

Citizen Band (CB) radio is another popular form of short-range radio communication during extended power outages and when cell phone and other forms of communication are down. Initially popular for truck-to-truck communications prior to cellular service during the 1970s, the CB radio suffered from too many people trying to talk on a limited number of channels. However, during a grid down

POWERING COMMUNICATIONS AFTER AN EMP

event it can be an excellent way of communicating with others beyond the limited range of FRS and GMRS walkie-talkies.

Fig. 19-3 4-watt 40 channel CB walkie-talkies.

A CB radio can be found in every tractor-trailer cab and all models are designed to run on 12-volt DC power and have rugged construction to withstand the constant highway vibration. Older CB radios were manufactured before the introduction of microelectronics and are considered to be less susceptible to EMP damage than newer models having integrated circuit devices, but the older units also have fewer channels. Keeping a CB radio in a metal EMP proof container until needed in an emergency can help guarantee survivability.

The FCC later expanded the original twenty-three channels to forty channels but lowered all new CB radios to a 4-watt maximum transmit power in AM mode, or 12-watts in single-sideband (SSB) mode. Since CB radio is still a very popular form of communication by truckers and RV owners traveling the Interstate, Channel 9 is reserved for emergency communications only. This channel is typically monitored by the state police who have a separate CB radio mounted in their cruisers in addition to their other communication radios. Channel 19 is reserved for non-emergency traffic news and updates concerning road conditions and warnings of accident backups. Whenever traveling cross country in our RV, our CB radio is always tuned to Channel 19.

POWERING COMMUNICATIONS AFTER AN EMP

Fig. 19-4. 12-volt DC powered 4-watt vehicle mounted CB radio.

During a grid down event, a vehicle-style CB radio and separate power supply can make an excellent "base station" for a rural home or off-grid cabin, paired with other family members and neighbors using handheld CB walkie-talkies. If you do decide to purchase a handheld CB radio pared with a CB base station, please note most of the detachable antennas supplied with these walkie-talkies are the short rubber ducky design. While these are less likely to break off when backpacking or walking perimeters, they will significantly reduce transmitter range. I suggest ordering the optional telescopic CB antenna which extends much longer and has the same quick disconnect as the short flex antenna that comes with each walkie-talkie.

Since most handheld CB walkie-talkies are supplied with separate 120-volt AC battery chargers, make sure you purchase the optional charger that will operate from a 12-volt DC vehicle dashboard outlet. All vehicle and base station CB radios are designed to operate from a 12-volt DC power supply, but during a grid down event you should have a larger RV/marine Group 27 or 31 size battery to power these higher wattage radios.

POWERING COMMUNICATIONS AFTER AN EMP

The private form of communications with the greatest range is ham radio. Depending on the license held by the operator, transmission power can be over 1,000-watts, with communication across continents and oceans depending on time of day and frequency. While learning Morse Code is no longer required, a written test is required, and most people will need to attend a weekly night class for several months to prepare using a well-organized workbook. Hams are very welcoming to all new members, and almost every city in the United States has ham radio volunteers teaching free classes several times each year. The most basic level of ham radio today is the Technician class license which is the first step for anyone wanting to become a ham radio operator. This license limits which bands you are allowed to use but does include the very popular 2-meter band.

Fig. 19-5. 80-watt 2-meter ham radio with separate 12-volt DC power supply.

POWERING COMMUNICATIONS AFTER AN EMP

Preppers have recently discovered the basic Technician class radio license will allow them to use a small 2-meter FM hand-held transceiver to communicate through a remote repeater station for crystal clear communication over a far greater distance than CB radios. Although still a line-of-sight form of communication due to the very-high frequency involved, there are 2-meter repeater towers located at high elevations near most cities and towns. When any handheld 2-meter radio is programmed with the correct "access codes" for a specific repeater, it will receive the weak transmission then instantly rebroadcast the signal using its much more powerful transmitter and antenna system. While any 2-meter radio can communicate directly with any other 2-meter radio that is within range without using a repeater, the range drops to only a few miles, while repeater-based communications can cover literally hundreds of square miles.[2]

While the Technician license and 2-meter radios may be the only level of emergency communications most preppers will need, there are additional levels of ham radio licensing which offer far more range and communication opportunities. The General amateur ham radio license requires a more thorough knowledge of radio and antenna theory but is still a relatively easy next step for anyone having a "Technician radio license.

The "General amateur radio license allows operating in many more frequency bands. Transmit power can be from 100-watts up to over 1,000-watts depending on band and type of transmission, so any prepper with this license and higher level of communication equipment will be able to talk with other amateur ham operators in every part of the world. Up to the minute news from distant countries can be obtained firsthand without being heavily edited or watered down by official news outlets. The ham radio bands are set aside for individuals, so business use and advertising are strictly prohibited.

POWERING COMMUNICATIONS AFTER AN EMP

Even the lower wattage ham transceivers can still communicate great distances and are designed for both base and mobile operations. A separate 120-volt AC power supply is required to provide 12-volts DC to power these larger power radios during normal times, so powering directly from a deep-cycle RV/marine battery is easy when in a grid down situation. While you may not see the need or have the desire to get into ham radio, at a minimum your emergency preparedness should include a quality battery-powered all-band radio receiver and a pair of GMRS walkie-talkies.

It is almost guaranteed that a high-altitude nuclear detonation will permanently disable all communication satellites in stationary orbit above the United States. In addition, damage to the solar arrays and stabilizing controls can also cause the satellites to fall out of position and eventually fall to earth. There are over a thousand satellites currently positioned above the United States that are an integral part of all phone, television, internet, and military communications, as well as land and marine GPS navigation, weather monitoring, surface mapping, drone control, and even military ground surveillance.

Certain layers of the ionosphere are used for shortwave radio propagation and reflecting radio waves back to earth which allows transmission far beyond the line-of-sight curve of the earth. No doubt a nuclear detonation that is designed to energize this ionosphere will temporarily alter or even block these radio transmissions. This should not last beyond a few days. Commercial AM and shortwave ham radio service should be quickly restored, although some forms of communication could be down for months or more. All of the battery-powered handheld radios and transceivers described in the chapter will survive an EMP or solar storm due to their small size, as long as they are not connected to a charger, mike cable, earphones, or extended antenna when an EMP hits.

CHAPTER 20

Powering Computers after an EMP

We all want to stay connected with the outside world, especially after an EMP attack or solar storm has caused a long-term grid down situation. Regular broadcast news if still on the air may be limited or controlled during extended emergency conditions. Being able to keep in touch using the internet to access alternative news outlets and emergency information will be a real-life saver, at least as long as the internet is operational. I'm sorry to tell you that tests have shown a regular desktop computer having a separate printer, keyboard, mouse, and monitor will not survive an EMP event. You will need to rely on a battery-powered portable computer device so make sure it will also contain the data you normally access from a desktop computer.

In my opinion, the greatest invention in home computers has been the laptop computer with longer-life rechargeable batteries. No longer do you need a television-size monitor and separate computer, keyboard, mouse, and printer requiring hundreds of watts to power and lots of cables. Recent advances in micro-electronics and battery technology have taken these already small laptop computers down to tablet and iPad size, while drastically increasing processor speed and battery life. Every laptop and tablet-size computer sold today includes either a built-in cell phone modem, wireless internet modem, or high-speed wired internet connection.

POWERING COMPUTERS AFTER AN EMP

Most people with a desktop computer system will usually have a uninterruptable power supply (UPS) with battery backup. Unlike a battery-powered laptop computer, a desktop computer will instantly loose whatever document you are working on if there is a disruption to the grid power, even if only for a fraction of a second. These UPS systems with internal battery will instantly transfer over to battery power the second grid power is lost to provide temporary 120-volt AC power for ten to fifteen minutes, depending on the size of your computer and monitor. The intention is to give you enough time to save your work, copy files to backup disks, and start an orderly power down of the system.

While not normally used to power laptop computers since they all have built-in batteries, a UPS system can provide substantial filtering and blocking large voltage spikes coming in on the house wiring. While not a true EMP protection device, these can significantly reduce the risk of EMP damage if you have several to each back up a separate laptop computer, flat screen television, DVD player, or CD sound system. In addition, more expensive UPS units have additional lightning type filters and larger batteries that can provide much longer runtime for these smaller electronic devices than possible when used to back up a high-wattage desktop computer.

While cell phone and satellite internet communications should still be available during the first few weeks of a grid down event, unless the outage was caused by an EMP, internet routers and satellite receivers will not work without grid power. Be sure the computer you plan to use during an extended power outage includes a way to access the internet without needing a home wireless router or satellite modem which are almost always 120-volts AC grid powered.

During the initial stages of a grid down event you may be able to keep a laptop or tablet computer fully charged from a generator while it is powering other loads, or from the 12-volt utility outlet in

POWERING COMPUTERS AFTER AN EMP

your car, assuming you have a 12-volt adapter and the right connecting cable. However, at some point the generator will run out of fuel and you may need to save any remaining gas in the car or truck for a possible evacuation later.

During an extended power outage, the easiest way to keep a laptop or tablet computer charged without a generator is with a foldup solar panel, which are available in 5-, 10-, 25-, and 40-watt or higher capacities. You will need a minimum of 25-watts to recharge most tablet computers in a reasonable time, and up to 40-watts for larger laptop computers. Even a small 10-watt solar charger can eventually recharge a larger laptop computer. However, it may take days of solar charging to provide one or two hours of computer use, so make sure you have at least one of the higher wattage solar panels to recharge any larger battery-powered devices.

Fig. 20-1. Foldup 25-watt solar panel will easily recharge a laptop computer.

POWERING COMPUTERS AFTER AN EMP

Although foldup solar panels over 25-watts in size can be fairly expensive, they will significantly reduce battery charging time, which is a real advantage on winter days with fewer sun-hours. A foldup solar charger is a must have item and can be easily packed and taken with you if you do need to bug out. Some of the newest foldup solar chargers are available with a built-in high-capacity battery. This allows you to charge the internal battery during the day when left in the sun, and then you can take it inside to recharge multiple small electronic devices at night or during periods of limited sunlight.

A battery-powered laptop or tablet computer with a built-in wireless internet or cell phone modem is a great way to keep in touch with friends and family by providing texting, e-mail, and public forum postings. Portable computing devices provide the ability to do an internet search on any topic, read about breaking news, movie reviews, access street maps, and get travel directions by typing just a few words in a search engine. You can instantly download radio show podcasts, instructional videos, and scan late breaking headlines from multiple news sources.

Even if you are operating on battery power and have multiple ways to keep the laptop or tablet computer fully charged, what happens if there is no internet available to access this information? A full grid down event may cause communication blackouts over multiple states and lasting months, and an EMP can actually damage or permanently destroy internet and satellite communication systems. Then what?

While things are still relatively normal, this is a perfect time to download reference books, manuals on homesteading and do-it-yourself skills, magazine archives, repair and parts manuals, building code books, emergency medical handbooks, pill and medical dictionaries, address lists, scanned insurance and loan documents, and anything you regularly use for your work or hobby. Most magazine articles can be downloaded and saved on your computer for reference later, and most publishers offer multiple-year anthologies of their magazines on CD in an easy to search format. There are thousands of classic books available to download for free due to expired copyrights.

Several publishers specializing in prepper products have scanned thousands of pages of survival and emergency backup power articles onto searchable memory sticks for access without the internet when attached to your laptop computer. I know, as I have found hundreds of my preparedness and solar magazine articles included in these collections without my permission, but that's another story! I suggest purchasing a quality encyclopedia in electronic format, which are typically available on CD or memory stick for use with a laptop computer. While not free, their cost is reasonable and will make a great way to research almost any topic without internet access or taking up several feet of bookshelves that you can't take with you if you do need to evacuate. Unfortunately, I have found several encyclopedias supplied on CDs that will *not* work without an internet connection, and this is not always noted in their advertising.

All mapping programs require internet access, but it's still possible to buy a good map data base that can be totally downloaded to a laptop computer or available preloaded on an optional memory card which will work with and without a GPS receiver or internet connection. There are many types to choose from based on geographic features. The cost is substantially less if you only need a detailed street or terrain map for a specific state instead of the entire country.

Some mapping programs for cell phones will provide exact GPS location and directions while walking along city streets, biking, or even hiking on an unknown mountain trail. However, most cell phone-based mapping programs only download the map section for your current location to save memory and time and require an internet connection to constantly download the map information for the area you are traveling into. Make sure the map software you purchase clearly indicates it will work totally offline using only the map data base that resides on your computing device.

If you do have to bug out and are traveling through areas with no cell or satellite internet service, being able to review a built-in map data base is a necessity, especially if you cannot access a foldout paper map or GPS receiver. Preplanning your bug-out travel should identify routes that avoid bridges, Interstate exit points, and congested cities that could become a major traffic bottleneck or police checkpoint during a crisis that would bring normal traffic to a complete standstill.

While you are downloading and storing all these reference materials onto your battery-powered computer, it's also a good idea to scan important documents that could be destroyed in a fire or flood. Homeowners and car insurance documentation, property deeds, financial and medical records, and banking and credit card information can all be safely stored using file encryption if your computer is lost or stolen. Irreplaceable family photos can also be saved and reprinted later if the originals are destroyed. Many big box stores offer do-it-yourself printing of photographs from a CD or memory stick, and the quality of these color photographs is equal to the original. While having a battery-powered laptop or tablet computer is critical during a major power outage, it will be even more useful if you can still access all of your important documents, maps, and reference materials when all communication systems are down.

Finally, even if you are able to keep your cell phone and GPS receiver powered up without a working electric grid, there is still the high probability that the satellites sending out the multiple GPS signals will be damaged or destroyed by an EMP. By now most people have trashed the foldout maps stuffed in the door pockets of their cars now that everyone is using GPS for directions. However,

POWERING COMPUTERS AFTER AN EMP

it is critical to keep a current foldout street level map for your nearest city, your state, and at least a multi-state highway map for your section of the county. I prefer the larger bound map books with each state on a separate page. If you are forced to leave your home or town heading out to a distant location, without a working internet or GPS system these old technology maps will be invaluable, and they are inexpensive.

CHAPTER 21

Powering Audio/Video Systems after an EMP

During an extended power outage and long after the grid and your generator have died, you will not have the power to operate large-screen televisions, CD and DVD players, or stream internet-based entertainment programs using devices that run on 120-volt AC power. In addition, if this is a true grid down or EMP event, all internet and normal communication systems may have stopped even if you have emergency power.

Thanks to the recreational vehicle industry and all the parents needing a way to keep the kids in the back seat quiet on a long drive, there are now available many sizes of battery-powered flat-screen televisions and DVD players. These are supplied with both a 120-volt AC charger and a separate plug to fit a 12-volt DC vehicle utility outlet.

Assuming broadcast television stations are still on the air even with a seriously reduced program schedule, these portable digital televisions can receive all television stations in range. They also come with a short antenna having a magnetic base and long cable allowing attachment to your vehicle roof for mobile use. If you live in a rural area, I recommend buying an extended-range digital broadcast antenna with signal booster, which requires 12-volt DC power to operate, but this will greatly extend the range and number of channels your battery-powered television can receive.

AUDIO/VIDEO SYSTEMS AFTER AN EMP

Fig. 21-1. Smaller battery-powered flat screen televisions have excellent reception, and most have inputs for DVD players.

Battery-powered DVD players are also popular for backseat kids' entertainment. A DVD player does not include a broadcast television tuner, but their built-in DVD drive and foldup color screen with stereo speakers are great for one or two people to watch a favorite movie when other forms of entertainment are without power. I recommend stopping by the discount DVD bin each week to build up a good supply of movies you can watch off-line, especially if you have young kids. I purchased several multiple sleeve DVD holders after discarding their plastic cases and can fit several hundred DVD's into each compact holder that are easy to grab along with a portable DVD player if you do need to evacuate.

Most of the portable DVD video players intended for vehicle or camping use have fairly small screens, but there are much larger flat-screen digital televisions sold for the mobile RV and boating

AUDIO/VIDEO SYSTEMS AFTER AN EMP

industry. These are especially nice during a grid down event as they are designed to run on 12-volt DC power. This means they are not only more rugged than a stationary home television, but they are also designed to be very energy efficient since they are battery powered and there are multiple ways to keep them charged.

There are several brands of dual voltage flat-screen televisions with removable bases and wide-screen format from 13-inch up to a 22-inch size. In addition to a broadcast television tuner, most include a built-in DVD player and inputs for playing videos or showing photos from memory cards and USB connected sources. The screen on the Axess 15-inch model I purchased was easily viewable by the whole family, and the built-in DVD player and stereo speakers only required connection to a 12-volt DC power source. When checking battery drain, I found the 15-inch model used half the power of the 22-inch model since battery life was my main concern.

If planning to bug out by yourself to a remote cabin, the smaller 12-volt DC personal-size television with DVD player will be the easiest to pack and requires the least amount of battery power. However, for two or more people, I recommend the 15-inch or 19-inch size dual-voltage flat-screen televisions having a build-in DVD player and stereo speakers.

If you are not an RV or boat owner, I strongly suggest checking out your nearest RV supply store. With the exception of the microwave oven and rooftop AC unit, every LED light fixture, kitchen appliance, water pump, furnace fan, exhaust fan, and entertainment device in any RV or boat is powered by a 12-volt deep-cycle battery. In addition, since audio/video equipment designed for RVs and boats are portable and battery powered, they are designed to be very rugged with a minimum battery drain. You can find the larger flat-screen televisions at any RV supply store, as well as 12-volt DC powered foldup portable satellite dishes for cable television reception anywhere.

AUDIO/VIDEO SYSTEMS AFTER AN EMP

Small E-readers or digital books are an amazing entertainment device and some models are available with color graphics screens. The Kindle Paperwhite e-reader isn't color but can operate up to two months on a single charge, depending on daily read time. These devices have built-in Wi-Fi modems with up to 8 GB of internal memory storage and are capable of holding literally thousands of books and magazines. All e-readers use a 5-volt mini-USB charger cable and can easily be recharged with a 10-watt foldout solar charger if you have the correct adaptors and cable. If left unplugged, their small size means very little chance of damage from an EMP.

Fig. 21-2. Battery-powered iPad and video projector.

AUDIO/VIDEO SYSTEMS AFTER AN EMP

The battery-powered video projector with built-in speakers shown in the photo will easily fit in the palm of your hand yet project a clear movie image up to ten feet wide in a slightly darkened room. While the built-in rechargeable batteries are limited to under two-hours of runtime, the projector has the ability to be powered from a larger external battery and recharged using a foldup solar panel. Most models I tested had multiple input jacks and cable connectors to project video from laptop computers, iPads, DVD players, and digital cameras. This combination of player and projector can entertain an entire room full of adults and kids, while requiring very little battery power.

Sitting by candlelight and playing card games, checkers, or working on a puzzle require absolutely no electrical power and can be very entertaining. There are plenty of board games you can have stored away for extended power outages or cabin fever when winter storms keep you home, and they fit nicely under the couch. This is especially important if you have younger kids to keep settled during what could be a very stressful experience.

All of the entertainment devices mentioned in this chapter should easily survive an EMP event due to their small size, as long as not left connected to any antenna or charging cable when an EMP hits. If you plan to keep these stored until actually needed, be sure to remove the rechargeable batteries. For any more expensive audio and video equipment, a better storage plan may be to keep these in their original packaging and use one of the EMP shielding methods described in chapter 36. Just remember all stores could be closed after an EMP or major grid down event. This means now is the time to stock up on the backup electronic devices, batteries, chargers, and things to keep kids occupied.

CHAPTER 22

Powering Medical Equipment after an EMP

While keeping the lights and a television operating during a major power outage makes life more enjoyable, assuming they were not damaged by an EMP, keeping medical devices powered up and special medicines refrigerated can be a matter of life and death. I have designed emergency backup systems for critical care patients living at home and some required an entire room full of electronic monitors, pumps, and refrigeration equipment.

Providing long-term backup power for a family member needing this serious level of home care is beyond the scope of this book. However, keeping a bedtime ventilator unit or heart monitor operating through the night is fairly simple, as these small loads can be powered all night by their own built-in battery system if you have a way to charge them up each day.

Fig. 22-1. 12-volt DC power BiPAP medical ventilator unit with backup battery attached.

MEDICAL EQUIPMENT AFTER AN EMP

Most of the bedside medical monitors and ventilating units I have tested require not only very little electrical power, but most are available with an optional 12-volt DC backup battery pack. A new BiPAP ventilating unit I recently tested consumed only 0.10 kWh of power at 120-volt AC (9 amp-hours @ 12-volt DC) for a typical eight-hour runtime per night. This would indicate a Group 31 size 12-volt RV/marine battery could easily power this unit for three or more days without totally discharging the battery.

The most reliable solution to power smaller medical devices is a totally separate battery backup system only connected to the medical equipment. A 12-volt AGM battery can be located near the medical equipment since there is no battery out-gassing from these sealed batteries under normal operating conditions. For larger power requirements, two 6-volt golf-cart AGM batteries wired in series to provide 12 volts are only slightly larger than a single 12-volt battery but have a much higher total amp-hour capacity and can easily handle the daily deep discharge and recharge cycling. Keeping a separate solar charger and battery just for powering medical devices will avoid having the battery drained by other non-critical electric loads in the home.

This isolated battery can be left connected to a 120-volt AC battery maintainer type 5-amp charger, which will keep the battery fully charged during normal grid operation and should recharge each day what the medical equipment discharges each night. The same charger can also be powered from a small generator during an extended power outage by just moving the power cord from the wall outlet to an extension cord supplied from the generator. A small 10-amp solar charge controller and 75-watt solar module is a good match for the daily load of this size battery and nightly power usage when both the electric grid and generator are no longer working.

If you require a wireless monitor that reports the status of a heart-pacer or other medical monitoring device, check with your doctor

MEDICAL EQUIPMENT AFTER AN EMP

to see if a dual-voltage model is available which allows powering from an external 12-volt DC source if the 120-volt AC grid or generator source goes down. Most of these devices must stay powered 24/7 which can be a significant drain on any backup power system. Chapter 26 will address how to keep things cold during a grid down event, which is critical if you have medications that must be kept refrigerated.

While all of us are accustomed to having a paramedic with a complete emergency room on wheels arrive at our door five minutes after dialing 911, that time is over after an EMP attack or major grid down event. There is no excuse for any able-bodied teenager or adult not knowing how to perform CPR or temporally stopping a major bleed. Part of any EMP preparations should include a significant upgrade in your emergency medical supplies and this does not mean one of the small plastic first aid kits containing nothing but a few band aids for minor cuts and some aspirin.

During a grid down event accidents increase significantly as there will be far more people outside lifting and removing storm damage, cutting firewood, climbing ladders to make roof repairs, and walking many miles if vehicles do not operate or roads are blocked. There are excellent books on emergency medical procedures and websites offering real emergency medical kits assembled by real medically trained professionals.[1] I strongly recommend keeping a good emergency medical reference library on your laptop computer and contacting the Red Cross or your local fire/rescue group to attend one of their free CPR training sessions offered multiple times each year.

CHAPTER 23

Powering Security Systems after an EMP

After a power outage lasting only days in a large city, and a few weeks in more rural areas, vandalism and theft will increase dramatically since the majority of any remaining residents still in any city or town will be without food and water. Local law enforcement will have far more to deal with than responding to a burglar alarm and may not show up at all. If you feel a gun is the best burglar alarm, keep in mind you have to sleep sometime, and many burglaries occur when the homeowner is home and asleep. A well-trained dog can make a great early warning system, especially since their exceptional hearing will raise the alarm long before someone reaches your door.

Increased vandalism during a power outage is also why many portable generators tend to walk away from garages in the middle of the night. Even a professionally installed security system and the communications needed to report alarms will fail sooner or later during a grid down event. We need a much simpler way to notify us when somebody is trying to break in or steal our property during an extended power outage and all grid-powered alarm systems have failed.

With the recent advances in wireless security cameras and motion sensor technology, there are very few homes or businesses today without at least some basic level of video security. While older security monitoring technology still requires a land line to report a fire or break-in, newer systems utilize cell phone and internet modems to send alarms to a distant emergency call center.

SECURITY SYSTEMS AFTER AN EMP

Unfortunately, while all security systems do include battery backup, most will fail when a power outage lasts longer than a week. In addition, during a true grid down event there may not be a local cellular or internet communication network still functioning that could report an alarm, assuming your security system is still powered up.

While there are all kinds of simple security alarm tricks including balancing a pop bottle on a doorknob or tin cans hanging from a string stretched across an outside walkway, there are portable battery-powered security devices that do not require any external phone or internet communication connection to function. Avoid the smaller models powered by button or odd size batteries which will be impossible to replace or recharge during a grid down event. You should be able to find models that operate on either AA or C size batteries. Later chapters will explain multiple ways to keep all battery-powered devices fully charged during a grid down event.

Portable door alarms the size of a cell phone typically include a large metal loop that hangs on any doorknob. Once the switch is activated, these devices provide an ear-splitting alarm if the knob is turned or the door is disturbed. Remote reporting motion sensor alarms can be used to guard a nearby garage or generator shed using a built-in motion sensor trigger. These are very similar in appearance to a typical wall-mounted motion sensor connected to a central alarm system. However, instead of an internal horn, these sensors activate a separate alarm receiver in your home using wireless technology with a range up to several hundred feet.

As mentioned previously, make sure to select a model that use either AA or C batteries which are the same size batteries used in all other battery-powered devices reviewed in this text. When it comes to home security, organizing an active neighborhood watch and using multiple walkie-talkies for communication is a great way to handle community security during a grid down event, while using battery-powered alarms to take care of all perimeter security needs. Chapter 19 provides more detail regarding battery-powered walkie-talkies.

For those living in more rural areas, there are several low-cost driveway and remote storage building alarm systems with both the sensor unit and separate alarm receiver powered by either AA or C batteries. By utilizing rechargeable batteries as discussed in chapter 29, these alarm systems can operate many months without any grid power. Several different alarm types are available, including a remote sensor that can be buried under or beside a driveway which only activates when a vehicle or large metal object passes by.

Other models are motion-based with the outdoor sensor located on a post or gate and will alarm when vehicles, large animals, or people pass near the sensor. Slightly more expensive units allow using one battery-powered alarm receiver with multiple remote sensors, and a different alarm sound can be selected for each remote sensor position making it much easier to identify the location of the disturbance. Due to their small size and no need for external power cables, these alarm devices will not be affected by an EMP.

Security studies for remote commercial facilities have determined that a totally dark building or equipment yard that suddenly lights up with motion-controlled security lighting is a better deterrent than leaving all lights on throughout the night. This will be especially true during a grid down event when entire towns may be dark, and all emergency power will be needed for interior illumination and not wasted on external lighting. This means the sudden and unexpected activation of perimeter lighting around your home or equipment storage shed when an intruder approaches will be a significant deterrent during this increased level of vandalism and break-ins that typically occur on dark nights during extended power outages.

Most homeowner's will already have one or more exterior floodlights around their garage and driveway that are motion

SECURITY SYSTEMS AFTER AN EMP

activated or use a dusk-to-dawn photocell control. However, none of these will be functioning during an extended power outage. There are many different sizes of battery-powered outdoor LED floodlights that are motion activated and include a solar panel to keep the battery fully charged. Solar floodlight kits having a higher wattage solar panel can power a DC floodlight containing multiple bright LED lamps and will easily illuminate a large parking area.

Unit controls typically offer different settings, such as dusk to dawn, motion sensor only, or a timer, to save battery charge. While you may already have multiple outdoor grid-powered floodlights, adding several motion-controlled LED flood lights that are totally solar powered could be a real lifesaver when all other exterior lighting fixtures have gone dark. While a dog barking can alert you to someone outside, it's safer if you can see what's causing the early barking! Solar-powered LED floodlights will operate no matter how long the power is out as they are always operating off the grid.

An important part of security, especially during a grid down event, is advance warning of any life-threatening events in your immediate area. During any type of major emergency, at least one AM radio station in each city is designated part of the National Emergency Broadcast System and will provide the same emergency information and evacuation advice to all affected areas. In addition, the National Weather Service (NWS) maintains a network of over a thousand automated weather radio stations, which cover all fifty states plus all coastal waters. These automated radio stations provide the first warnings of any severe storms approaching your area and recommend actions to take.

The National Weather Service operates on seven specific frequencies in the 162.40 to 162.55 MHz high-frequency band, but these are outside the normal frequency range of any household AM or FM radio. You will need to purchase a specific NWS radio to receive these weather alerts, but these radios are small and very

SECURITY SYSTEMS AFTER AN EMP

inexpensive. Battery-powered NWS radios are also easy to find. Although they have to be turned on at all times to receive the alerts, this is not a nuisance since just before any warning a special tone is sent which activates the speaker in the receiver. Normally these radios make no sound at all.

NWS weather alerts can also be sent directly to any e-mail address and to any cell phone as a text message if you register on the NWS website. There is no cost for this service and a test signal is sent every Wednesday afternoon at two specific times so everyone can verify their radio is still operating properly. Having a battery-powered NWS radio can be a real-life saver during any weather emergency, especially if you live in an area that experiences frequent tornadoes.

Normal weather updates for your area are broadcast on the hour and eleven minutes after each hour, but you can mute your NWS radio to stay quiet until it receives the activation code. When the activation code is received, your NWS radio's speaker is turned on and you will then hear an automated message describing the potential hazard. In addition to warnings for severe storms, hurricanes, and tornadoes, this same system is also used to notify residents about potential flooding, forest fires, dangerous ocean or river conditions, unusual solar activity that could disrupt communications, and even missing child AMBER alerts.

Each county of each state has an assigned code, and each type of alarm condition is also assigned a specific code. When you purchase a NWS radio, you can indicate which counties are associated with your area so your NWS radio will only activate when a warning is associated with your immediate area. In addition, you can also select which types of warnings you wish to receive.

SECURITY SYSTEMS AFTER AN EMP

Fig. 23-1. Battery-powered NWS emergency radio with both car charger and 120-volt AC wall charger.

While most NWS radios are tabletop design and include an alarm clock function as well as a basic AM/FM radio, I recommend for grid down concerns one of the portable battery-powered NWS radios. While these include a grid-connected power adaptor and charger base, you can take the radio with you if you do need to evacuate and they do not need the other power draining functions normally included with the tabletop models. When removed from the charger, it switches to internal battery power and will then start automatically scanning for the closest NWS radio station along your route. I really like the "Midland HH54" and the "Oregon Scientific WR602" portable weather radios, as both have long battery life when used out of the charger and both use AA size rechargeable batteries.

SECURITY SYSTEMS AFTER AN EMP

I am sure by now everyone has one or more battery-powered smoke detectors in their homes, apartments, and businesses. Not only do they really save lives, but they are very inexpensive and easy to install. While most will operate up to a year on the internal battery, its good practice to change the battery when you change the clocks to daylight savings time which serves as an easy reminder. The smoke detector is just another battery-powered device we depend on each day to protect our lives, but we still need fresh batteries when it's time to change and these may not be available during a grid down event.

You can also find battery-powered carbon monoxide (CO_2) and natural gas/propane detectors which look almost identical to a battery-powered smoke detector. If you have a basement or mechanical room that includes gas appliances, these are a great early warning system. These self-contained alarms are too small to be affected by an EMP event, especially since they are not connected to any power wires.

In many larger homes and apartment buildings the smoke detectors you see in the ceiling may not be battery powered. In fact, many are now powered by 120-volt AC grid power and report an alarm condition back to a central alarm panel which is also powered by the grid. While these central fire alarm panels do have battery backup, the batteries are not sized to power these systems for an extended power outage, and without grid power the systems will soon stop providing fire alarm protection. If you find yourself living where this is the case, I suggest purchasing several low-cost smoke detectors that do not depend on grid power to operate.

Another area we take for granted is telling time. No doubt you have a digital wristwatch, bedside alarm clock, or microwave oven clock. You also have a small clock display on your cell phone and laptop computer, and hourly radio time announcements. In fact, it's almost impossible to not know the time today. However, most of these clock devices require the utility grid for power or to recharge. During a real grid down event, most of these devices will fail or just stop working, and the simple act of knowing the time may not be so simple.

Almost every office supply and furniture store sells large-dial wall clocks that are very inexpensive. Unlike all of the other clocks you own, these wall clocks are normally powered by one or two AA size batteries and will run up to a year before needing to replace the batteries. Having at least one large wall clock in the house that keeps perfect time without the utility grid or internet is another essential low-cost preparedness item. Since the actual battery-powered mechanism is smaller than a pack of cigarettes, regardless of how big the clock face is, all of these clocks are EMP proof.

While you are shopping for a battery-powered smoke detector and wall clock, this will be a good time to also purchase several A-B-C type fire extinguishers for the kitchen, garage, and each vehicle, and be sure their location is visible and easy to access. Working for years with our county's volunteer rescue squad, I have seen many small kitchen and vehicle engine fires that could have been easily extinguished with a small handheld fire extinguisher if the owners had actually owned one. Instead, these small fires became big fires in mere minutes and soon engulfed their homes and vehicles.

During a grid down event, your local fire department and rescue squad will no longer be just minutes away, and you could be on your own instead of the fast emergency help you are accustomed to receiving, so better prepare now.

When reviewing the many ways to improve your security with battery power, I hope you realize security will involve more than smoke detectors, motion alarms, and checking your door locks. Take time to review the types of hazards your home faces, and make sure you have a way to minimize each risk. This is also a good time to start a neighborhood watch program if you do not have one already.

SECURITY SYSTEMS AFTER AN EMP

This chapter covered many different types of security concerns related to living through a grid down period or after an EMP attack, and they are all important to consider. The battery-powered security devices I recommend have followed the other chapters advice by standardizing on the AA and C size rechargeable batteries and all are EMP proof. When it comes to security, remember, sometimes things just do not go the way you planned so always have a backup.

CHAPTER 24

Protecting Battery-Powered Tools from an EMP

During a long-term power outage or following an EMP attack, it is doubtful you will need to build a house! However, you may need to repair storm damage to an existing house, board up windows or service a vehicle. You may have a garage full of 120-volt AC power tools. However, these will be useless during a true grid down event, especially after your generator has run out of fuel or was damaged by an EMP.

I strongly recommend buying a complete set of portable battery-powered tools that use the same size and type battery. There are several name-brand suppliers offering cordless power saws, drills, work lights, and cut-off saws, and all use the same rechargeable batteries. If funds are limited, you can't go wrong with a battery-powered Sawzall. These saws are amazingly versatile and have different specialty blades for cutting wood, metal, and even blades that can cut wood having embedded nails for demolition work after a storm. As long as they are not actually sitting in a charger base, most battery-powered tools will not be damaged by an EMP event. Storing them without a long blade or drill bit will reduce the potential "antenna" effect that can absorb E1 energy.

I served several years with a local volunteer rescue squad, and we had a battery-powered Sawzall on every rescue vehicle and fire truck. This should be your first battery-powered tool to start your collection. The newest models are now designed to use a 20-volt

PROTECTING BATTERY-POWERED TOOLS

lithium-ion battery, which is an ideal size for both cutting power and battery life. Your second battery-powered tool should be a multi-speed 1/2-inch capacity drill. In addition to all sizes of drill bits, there are screwdrivers, wire brushes, and even paint stirrers to fit these drills. There are also small water pumps available to fit a battery-powered drill which can be connected to a garden hose for both filling water storage containers and removing standing storm water.

Fig. 24-1. Many different types of battery-powered tools are available, and all can us the same 20-volt batteries and both AC and DC chargers.

Since many manufacturers are converting over to longer run-time lithium-ion batteries for their cordless tools, they now offer battery-powered chainsaws, which actually work reasonably well when using these higher-capacity batteries. A battery-powered chainsaw stored in your vehicle can be a real lifesaver for clearing downed trees and limbs blocking your evacuation route, and since no gas is required, they can easily store behind the seat.

Most brands and sizes of chainsaws, regardless of being gasoline or battery powered, tend to leak bar oil during storage which can ruin vehicle carpets or upholstery. To avoid this mess, you may want to leave the automatic oiler tank empty and instead apply chain oil

PROTECTING BATTERY-POWERED TOOLS

using a separate squirt bottle to spray oil directly onto the chain when cutting during an emergency.

Most battery tool purchases usually come with their own 120-volt AC battery charger. Without grid or generator power you will have no way to recharge these power tools once they are discharged. While not normally found in a typical builder supply outlet, battery-powered tool manufacturers offer a separate charger that plugs into your car's 12-volt DC dashboard outlet and does not require the grid or a generator to operate.

You will need to do some internet searching to find a supplier as these are not normally available locally. It is very important to have several battery-powered tools and a way to keep them charged during a grid down event. In addition, since these special chargers are designed to operate from a 12-volt DC vehicle outlet, they can also be powered from a separate deep-cycle 12-volt battery that is kept charged using a solar panel.

It's important to standardize on a single tool brand and battery voltage so you can use a common 120-volt AC charger and 12-volt DC charger to keep these operating during both normal every day and grid down times. Another advantage of standardizing on a single brand is you will not need a separate size or type of battery for each power tool.

Having a primary battery and one or two spare batteries that can be charged while a tool is in use makes it easy to always have several fully charged batteries that can be shared with whatever tool is needed at a given time. This also avoids the problem of having a battery staying in each separate tool and they all are totally discharged due to minimal use of each tool. Always try to find the optional higher-capacity batteries when buying spares.

While you do not want to overload your source of backup power, most brands of chargers for battery-powered tools offer both standard and fast chargers. Although the fast charger draws more

PROTECTING BATTERY-POWERED TOOLS

amps, in some cases they can cut your charge time in half. This is a real advantage if you are powering the tool charger from a generator or vehicle dashboard outlet as this could cut fuel usage by more than half for a typical battery charging cycle.

As I discussed in far more detail in my last book, *Lights On*, when living through a grid down event, regardless of the cause, most of the appliances and electronic devices still operating will be battery-powered. However, unlike 120-volt AC appliances and having multiple wall outlets in every room, you may be limited to a single 12-volt dashboard outlet in your vehicle and perhaps only one or two foldup solar panels.

You will need a multiple 12-volt adaptor for each vehicle which allows plugging up to three "cigar" type charging adaptor plugs into a single 12-volt dashboard outlet. You may also want to add a multiple USB adaptor which provides five or more 5-volt USB outlets from a single 12-volt dashboard outlet. Afterall, during an extended power outage, assuming your vehicle still runs and you have fuel, you do not want to run the engine for four hours just to recharge a single cell phone!

CHAPTER 25

Powering Water Pumps after an EMP

I am sure by now everyone knows the importance of having an emergency supply of food and water as backup for an extended grid down event. Unfortunately, even a barn full of water barrels sooner or later will become empty during a real grid down disaster since without fuel your generator cannot power your well pump to refill them. You need a simple and foolproof backup plan to have clean water, even if this means no running water in the house.

For thousands of years civilizations relied on lifting or pumping well water or river water to a higher elevation by hand or animal power, then carried it by hand wherever needed using buckets or clay pots. While a standard toilet can be flushed indefinitely using a two or three-gallon bucket of water carried from a nearby lake, creek, or even a swimming pool, having clean drinking water requires a totally different solution.

There are several easy ways to obtain clean water from an existing well when the grid is down. My two-hundred-foot-drilled well has a standard 240-volt AC well pump located near the bottom which is piped into the house. It is normally powered by the utility grid, but if the grid is down it can be powered by my solar power system or my whole-house generator during a power outage. I also have a smaller capacity DC well pump located in the same well but positioned slightly above the AC pump. Both pumps are piped together using check valves at the point where the separate supply

POWERING WATER PUMPS AFTER AN EMP

pipes exit the well casing. This avoids one pump trying to pump back into the piping of the other pump.

My pressure tank has two pressure switches, with the highest-pressure setting to control the AC pump, and the lowest setting switch to control the DC pump. If both grid power and generator power are lost, the AC pump will stop pumping and the water pressure will continue to drop until the switch controlling the DC pump is energized. Although the DC pump is much smaller and pumps less than half the flow and pressure of the AC pump, it is always available since it is powered directly from my solar-charged battery bank and does not require an inverter to convert battery power to 120-volt AC power. I also installed the largest pressure tank I could find which is about the size of a hot water tank. The large storage capacity significantly reduces the constant cycling of either pump. Constantly starting and stopping a well pump is hard on both the pump and the inverter or generator powering the pump and will shorten their operating life.

If your well is not close to the house or you have a different well located at some distance from your home or week-end cabin, another solution may be to use a DC submersible pump as the only pump in the well. It will be powered from a nearby pole-mounted solar panel which keeps the solar panel and wiring up and away from potential animal damage. This system does not require any batteries and the operation is very basic. This is also a very common method used on today's larger farms and ranches for watering cattle in distant fields where there is little surface water.

When the sun is up, the pump starts pumping, and when the sun goes down, the pump stops pumping. Of course, you will need someplace for this water to go as even a small three-gallon-per-minute solar pump can move over twelve hundred gallons in a single day. Typically, this water is piped directly into a concrete or galvanized-steel storage tank located at a higher elevation than the house or cabin.

POWERING WATER PUMPS AFTER AN EMP

Fig. 25-1. Solar-powered water pump with pump controller next to well.

For every 2.3 feet of elevation, you will get an additional one pound per square inch (psi) of water pressure, so to achieve a typical city water pressure (30 to 40 psi) in the lower house, the water storage tank will need to be located seventy to ninety feet higher in elevation. Keep in mind this is a measurement of the change in *elevation*, not a measurement of the *distance* between your home and the tank. In northern climates the tank may need to be below ground unless it is several thousand gallons in size and the water flows in and out is fairly constant to prevent freezing.

You can wire the pump directly to the solar panel without a solar charge controller or batteries and it will work. However, most solar pump installations usually include an optional pump controller which will improve the pumping and protect the pump when its connected between the wires from the pump and wires from the solar panel. During the first and last hours of daylight, the solar panel may still be producing an acceptable voltage, but not enough amp flow to start the pump or keep it pumping. Without the controller, the pump will stall and stop pumping during these low light periods but will still have some electrical power passing through the pump's stalled motor. Over time, this can cause the motor's windings to overheat and eventually short out or dry out the lubrication in the motor bearings.

POWERING WATER PUMPS AFTER AN EMP

The pump controller blocks all solar power from going to the pump until it is within the voltage and current range for the pump to run safely. In addition, more expensive pump controllers are able to convert any excess voltage into additional amp current during these lower sun hour periods, which can power the pump at slower speeds when it would normally not have enough current to run at all.

Most pump controllers also include replaceable fuses to protect the wiring and added terminals that allow connecting float switches. These typically look like a rubber softball and contain a waterproof switch and several feet of waterproof cable. These can be installed in the tank to start and stop the pump based on water level in the tank. They can also be installed in the well to stop the pump if the water level in the well gets too low. While the submerged solar well pump may survive an EMP event if well grounded, it is doubtful the pump controller will survive. Although small, the long wires will absorb the EMP shockwave and damage the controller's internal electronics. Adding ferrites to the wires going into and out of the charge controller will help reduce this risk.

For those seeking a less technical solution, there are several high-quality hand pumps now available that attach onto the top of a standard four- or six-inch well casing. These are not the old cast-iron pitcher pumps with leather plungers you see in photos of farmhouse life during the 1930s. These have all stainless-steel and brass construction and will provide many years of service. These new hand pumps are easy to install since their dip tube and check valve easily extend down into the well past any existing power wiring or piping for the AC powered pump located further down. While these take some exercise to operate, hand pumps are guaranteed to be EMP proof!

Without grid power, once a generator is out of fuel there will probably be no running water in the house unless you have a solar-

POWERING WATER PUMPS AFTER AN EMP

powered pump. After an EMP event, the grid could be down for many months and the generator could be out of fuel or its controls damaged. This also means no functioning toilets unless a nearby pond or creek provides flushing using a hand-carried water bucket. However, a compost toilet is a great backup plan as they require no sewer connection and no water plumbing, so they can be located almost anywhere.

Most models have a small 12-volt DC solar-powered exhaust fan which pulls air from the toilet and exhausts to the outside eliminating all indoor odor. Camping outlets also offer a toilet seat with fold-down lid and disposable bags that will convert any five-gallon plastic bucket into a toilet after adding a little peat moss.

During a grid down event, clean drinking water may not be available regardless if you are on a city municipal or private water system. Clean drinking water is also a concern if you must evacuate from your home or bug out with only what you can carry. We use a large Berkey water filter for our home, and a smaller model for our truck camper. These filters will remove all contaminants in the water used to fill the top tank, including bacteria, parasites, cysts, lead, and mercury, and the filter elements last many months.

Without an alternate way to power a well pump, bathing will be mostly by wet wash cloth. However, there are battery-powered pumps the size of a pear with six-foot hose and shower head attached. By dropping into a bucket of water, hopefully heated on a propane grille or wood fire, and then placing the bucket in

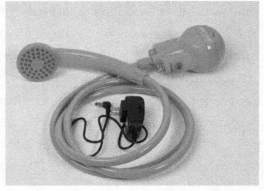

Fig. 25-2. Waterproof battery-powered bucket pump and shower head makes a great shower when the grid is down.

POWERING WATER PUMPS AFTER AN EMP

your shower stall, these will easily provide a warm shower lasting five minutes. These have a waterproof USB type connector and cost under thirty-five dollars. Being able to provide your family's drinking water each day from virtually any surface-water source means you will not be standing in line for hours with everyone else waiting on a FEMA truck to hand out a few disposable bottles of drinking water. One of the first things that happens in any disaster is the local population becomes sick from drinking contaminated water. Having a tabletop water purifier at home and a water bottle with filter for the road or hiking are a must have.

CHAPTER 26

Powering Refrigeration after an EMP

If a power outage lasts longer than several weeks or you experience a real EMP event, you will probably have already eaten everything in your freezer, even if you have an alternate way to power and it was not damaged by an EMP. Grocery stores will be closed and shelves bare, and with nothing to refrigerate, I'm afraid trying to power a conventional refrigerator or freezer at this point will be a waste of effort and scarce power resources.

History has shown time and again that meals and meal preparation will become a much simpler affair after all of your kitchen appliances stop working, the store shelves are empty, and there is no water coming out of the sink faucet to prepare meals and wash dishes. The primary meal during a true grid down event will be a large pot of soup or stew, with an attempt to provide some variety each day. No more taking a frozen chicken or beef roast from the freezer, turning on the oven timer, and setting the table.

The sobering fact is, most of us will have a very empty freezer during a true grid down event even if we have a generator and large fuel supply to keep it powered. A kitchen refrigerator is one of the largest power consuming appliances in a home and keeping one operating during an extended power outage requires far more backup power and effort than you would think, assuming your generator still has fuel.

REFRIGERATION AFTER AN EMP

Refrigerators and freezers do not require continuous electrical power, but they do cycle on for ten to twenty-minutes each hour depending on the ambient temperature. This means electrical power will need to be immediately available every time they cycle. Most new refrigerators are better insulated and should be able to hold their temperature through the night if their doors are kept closed, but they will need to cycle on again first thing in the morning to make up for any internal temperature increase during the night. Early morning is the time family members will be getting up and wanting to bathe and eat, even during a grid down event.

Manually starting a generator during the early morning, if it's still operational, will maximize generator use by powering the well pump, cool down a refrigerator, and power the coffee pot, microwave oven, hairdryer, and room lights since these will all need to operate at the same time when getting up and ready for breakfast. Running a generator for two hours in the morning should provide enough run time to cool down the refrigerator and freezer while everyone is getting up, bathing, preparing breakfast, and eating.

After breakfast the sun should be up and providing plenty of daylight in most interior areas of the home or cabin, and the refrigerator should be cooled back down enough to coast until evening if the door is kept closed. Manually starting the generator again for two hours during evening meal preparation when the well pump, room lights, microwave oven, and other kitchen appliances are again needed will allow the refrigerator time to cool back down again.

By concentrating electrical power needs to just the two periods of breakfast and dinner, this schedule will keep your generator run time to a minimum while still keeping the refrigerator reasonably cold. During a real grid down event, supper may become the main prepared meal, with breakfast and lunch consisting of pre-packaged snacks or foods that do not require refrigeration or cooking.

REFRIGERATION AFTER AN EMP

All modern refrigerators and freezers use an electronic control board to maintain temperatures and cycle the compressor. In fact, some models have more than one electronic control board located in multiple locations and EMP testing has shown their microchip components will not survive an EMP.

While it is expected the metal refrigerator housing will help attenuate the high-frequency E1 energy reaching the unit through the air, the E2 and E3 energy surge in the house wiring will easily damage the electronic boards even if the heavier gauge wiring in the compressor does not suffer damage. This means there needs to be other ways to have at least some refrigeration for easily spoiled items like milk and butter, and especially any medications that require constant refrigeration.

What if you could make ice and store it in a high-quality ice chest? You could easily get by indefinitely without a large refrigerator or freezer and would not need the large generator it takes to cool one down each and every day. Making ice is not only easy without a working freezer, it's also very fast. I can absolutely guarantee that during an extended power outage there is a way to make ice cubes in seven minutes and at room temperature, while requiring very little electrical power. You can keep this up all day as long as you have a supply of clean water, a small inverter, and a fully charged deep-cycle RV/boat size battery. In addition, this battery can be kept fully charged using your generator, your car or truck, or even a single solar panel.

I found a wide selection of portable countertop ice makers that are perfect for this application costing in the $120.00 to $150.00 range, depending on design features. These are small enough to survive an EMP attack if kept unplugged when not in use. The only downside with each of these lower cost ice makers is you have to fill the water reservoir by hand and the ice cube compartment is not refrigerated, so the ice will melt if left in the unit. When we pick a

REFRIGERATION AFTER AN EMP

day to make ice, we check the ice maker twice each hour to empty the hopper full of ice cubes and put them in our freezer.

During a full day we can fill two separate ten-pound bags of ice which easily lasts us up to two weeks. Being able to make an unlimited supply of ice during a power outage lasting months without a working refrigerator or freezer is like magic. In addition, unlike most of the emergency equipment many people buy then store away for a major disaster that may never come, a portable ice maker can be used regularly during normal times which is a real advantage.

I tested both the hOmelabs, #HZB-12/A and the VREMI #VRM010636N ice makers and found either would be a lifesaver after a grid down event. Both ice makers retail for $159.00 and have a similar size and weight, and both make the same size ice cubes.[1] I wanted to be able to power an ice maker from multiple sources during a major grid down event and without a large generator.

By adding a small 12-volt DC to 120-volt AC inverter, these ice makers can be powered all day from a single 12-volt RV/marine battery, which in turn can be recharged at the same time using a solar panel. In addition, when not making ice, this same battery and solar panel can also power your TV or satellite system which also typically require 120-volt AC.

Fig. 26-1. Tabletop VREMI portable ice maker can make ice cubes in just seven minutes.

REFRIGERATION AFTER AN EMP

Most ice makers this size have a very-high amp draw for one to two seconds during the start of each ice making cycle which occurs every seven minutes, and then settles down to a constant 120-watt draw for the remainder of each cycle. I found a pure sinewave 800-watt inverter would easily handle this brief in-rush current peak at the start of each new ice making cycle, but several 400 to 600-watt modified sinewave inverters I tested could not handle this high in-rush current and would shut off on overload. An inverter this small should be able to survive an EMP event, as long as it remains disconnected from any power cords or battery wiring when not being used.

While keeping the ice maker unplugged or stored when not needed will prevent E2- and E3-induced high-voltage from entering through the house wiring, E1-induced high-voltage spikes can still energize the long power cord of the unit even if rolled up and left unplugged. While there are no long wire runs inside these ice makers to serve as an antenna to absorb E1 energy, shortening the power cord to less than eighteen inches will provide extra insurance. While cutting the cord and installing a shorter three-conductor plug is fairly easy, if you are not comfortable doing this type of wiring, perhaps you have an electrician friend who can help as this would take only a few minutes.

A typical group 31 RV/boat battery will have an average charge capacity of 110-amp hours, or 55-amp hours by staying above a 50 percent charge level to extend battery life. I can operate either ice maker for about four hours while keeping above this 50 percent low battery limit. During a true grid down event, you will not want to just run the ice maker four hours and then stop. You will want to be charging the battery that is powering the ice maker at the same time you are making ice.

To offset the 120-watt constant load of the ice maker requires a 125 to 150-watt solar panel, which is still far less solar power than

REFRIGERATION AFTER AN EMP

needed to keep a standard residential-size refrigerator/freezer operating twenty-four hours per day. Obviously, this means making ice when the sun is shining! These ice makers can also be powered by your car or truck battery while the engine is running. Any inverter powering a load over 300 watts will draw too much power to use the 12-volt outlet on your vehicle's dashboard. An inverter sized to power an ice maker will need to connect directly to the posts of the vehicle's battery using the recommended size battery cables and keeping these as short as possible.

While I am currently putting the ice cubes into our grid-powered freezer, after a grid down event, I recommend storing the ice in a small super-efficient ice chest having a gasketed lid and thick wall insulation. These can keep your ice cubes frozen for days. Don't pick a unit that is too big, as you are just trying to keep ten pounds of ice frozen. An oversized ice chest will cause the ice to melt faster.

When we need less ice, we usually fill two or three wide-mouth thermos bottles, which really keep the ice cubes frozen for many days, and this is much more convenient when camping. Being able to power the ice maker from the utility grid, your backup generator, your car or truck, or a solar charged battery provides multiple ways to make an endless supply of ice cubes during a true grid down event, and this size ice maker can go with you if you are forced to bug out.

If you have a vehicle that is still operational and have lots of fuel long after an EMP or other grid down event, there is another way to have some limited refrigeration. There are now high-quality refrigerators and freezers that are not much bigger than a microwave oven, which are intended for in-vehicle use. Some larger models have a separate freezer compartment, but most operate as a refrigerator only or freezer only, depending on which temperature setpoint you use. These are small enough to be powered by the 12-volt DC dashboard outlet in any vehicle.

REFRIGERATION AFTER AN EMP

Most inexpensive portable battery-powered refrigerators typically use a semiconductor device that gets hot on one side and cool on the other side. Unfortunately, this low-cost technology cannot lower the temperature enough to freeze anything and will barely cool a drink on a hot day. The portable refrigerators I am describing in this chapter use the same type of motor-driven refrigeration compressor as found in your kitchen refrigerator, and some models can easily reach below zero temperatures.

To prevent over discharging the vehicle's battery, you should only power these units when the engine is running which usually means while driving. However, it is easy to power these from a separate 12-volt sealed Group 31 RV/boat AGM battery, and recharge this spare battery using a generator, the grid when available, or even a 100-watt solar panel and 10-amp solar charger during a real grid down event.

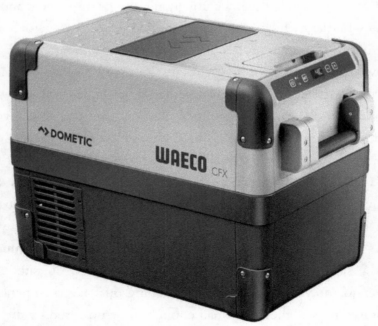

Fig. 26-2. Dometic portable refrigerator/freezer operates on 120-volt AC or 12-volt DC.

REFRIGERATION AFTER AN EMP

I own the Dometic #CFX28 portable refrigerator/freezer which costs $650.00 and has a 27-quart capacity or can hold about forty-eight canned drinks. I like this model as it includes both a detachable 120-volt AC power cord and a detachable 12-volt DC cord and works the same from either source of power. It's also EMP resistant as both power cords detach from the unit leaving no external wires to serve as an antenna to absorb E1 energy, and if unplugged from either power source will also not be impacted by E2 or E3 high-voltage surge that normally are carried by any power wires.

For those wanting a much simpler backup refrigeration solution, Dometic offers an optional battery pack about the size of a small shoebox. It has very rugged construction and foldup handle making it easy to go portable. The lithium iron phosphate battery stores forty amp hours of power, which provides almost forty hours of stand-alone refrigerator/freezer runtime and should work with any portable 12-volt DC refrigerators. The battery pack includes a digital battery meter, built-in charger, extra USB outlets, and sockets to connect a solar panel, generator, or even grid power when available for recharging.

The WHYNTER #FM-45 G is very similar in design features as my Dometic unit but holds up to sixty canned drinks and costs two hundred dollars less. The smaller ALPICOOL #22 only holds thirty canned drinks and will only operate on 12-volt DC, so it does not include the 120-volt DC power cord. However, this unit costs less than half the other models and still has hundreds of positive customer reviews.

All three models will easily reach -4°F, which is well below freezing when switched from refrigerator to freezer mode. Since any of these portable refrigerators would be great to take camping or on long road trips, they will have multiple opportunities for use during normal times while always ready for a future grid down event.

CHAPTER 27

Using Vehicles for Battery Charging

Not everyone owns a large whole-house generator for backup power during a power outage, or do they? That car or truck sitting in the driveway is basically a $20,000.00 portable generator, and the early EMP vehicle testing indicated some will still be operational after an EMP, while others may require some repairs as discussed in chapter 11 but can be made operational again.

There are two ways you can turn any vehicle into a source of backup power. For smaller 120-volt AC loads inside the house, a 12-volt DC to 120-Volt AC inverter up to 150-watts are available to plug directly into the 12-volt dashboard outlet. However, for larger loads, an inverter up to 2,000-watt capacity can be connected directly to the vehicle's battery terminals using the heavy-duty cables provided.

Once connected to a quality outdoor rated extension cord, it's easy to supply multiple 120-volt AC appliances and cool

Fig. 27-1. Portable 150-watt 120-volt AC inverter plugs into any 12-volt DC vehicle dashboard outlet.

USING VEHICLES FOR BATTERY CHARGING

down the freezer, while powering all kinds of small electrical appliances, and simultaneously recharging the batteries in multiple cell phones, laptop computers, and communication radios in the house. You can use the 12-volt DC dashboard outlet in a vehicle to recharge multiple DC powered electronic devices, assuming you have the optional 12-volt DC charger and charging cable for each specific device.

It is very important how you use a vehicle to power electrical appliances or recharge portable devices in order to minimize the fuel usage. While this does not mean trying to power everything in your house to the point of blowing fuses or melting extension cords, it also doesn't mean running a car engine for six hours to recharge a single cell phone!

While almost everyone has multiple-outlet extension cords to power 120-volt AC appliances, not everyone has similar extension cords to allow charging multiple 12-volt DC devices at the same time. All types of 12-volt DC multi-outlet adaptors can be found at most RV and boating supply outlets. In addition, most 12-volt charger cables for cell phones, laptop computers, and portable entertainment devices have a USB type plug end.

There are now adaptors that have a single plug to fit the 12-volt dashboard in any vehicle, which contain multiple USB type outlets. This allows charging four or five cell phones, laptop computers, and electronic devices at the same time, which would significantly reduce a vehicle's engine run time and fuel usage. Most 12-volt dashboard outlets in a vehicle are supplied from a 15-amp DC fuse and wiring, which provides over 150-watts of charging power.

A heavy-duty truck battery usually has thicker lead plates which can withstand a daily deep charge/discharge cycle and can power a constant load for long hours. However, most smaller car batteries have very thin lead plates to reduce weight and only need to provide

USING VEHICLES FOR BATTERY CHARGING

one large discharge during starting, then a long period of slowly recharging as the vehicle is being driven.

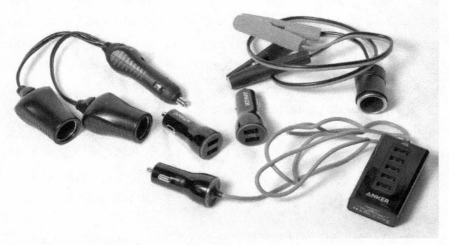

Fig. 27-2. Multi-outlet 12-volt adaptors allow charging multiple 12-volt DC and multiple USB powered devices from a single vehicle dashboard outlet.

Charging any portable device using the vehicle's 12-volt utility outlet can quickly discharge a smaller car's battery unless the engine is running, and most vehicle alternators do not generate any extra charging amps unless the engine is operating above its normal idle speed. While trucks typically have both a heavier battery and alternator, there is still the danger of discharging the battery below its ability to restart the engine when using the 12-volt utility outlet for any length of time while the engine is off, especially when trying to power a large electrical load.

During a grid down event, it is recommended to park the vehicle as close to the home as possible, and the extension cord from the inverter to the appliances in the home should be at least a 12-gauge wire labeled for exterior use. These heavier outdoor extension cords are typically yellow in color and will have a much lower voltage drop

USING VEHICLES FOR BATTERY CHARGING

than indoor-type extension cords, especially if one hundred feet long.

Unfortunately, you will probably need to keep the vehicle's hood up during any temporary use of the vehicle to power household electrical loads as inverters this size will need to connect directly to the vehicle's battery posts using very short heavy-duty cables. Since more than one inverter and its cables have been eaten by engine cooling fans, mounting the inverter on a large flat board is recommended!

Some preppers mount an inverter inside the truck cab and use permanent heavy-duty battery terminals and cables to make the electrical connection to the battery. I have a 1,500-watt inverter mounted behind my truck seat along with a separate heavy-duty extension cord, electric chainsaw, and portable light. If I need to remove a tree blocking the road after a storm, I never need to worry about a gasoline chainsaw that leaks oil and gas or will not start, or a battery-powered chainsaw that runs out of charge before the cutting is done. Being prepared means never letting your vehicle's gas tank drop below halfway, especially during the winter months.

CHAPTER 28

Charging Batteries with Solar Power

Most of this text describes how solar power should be a part of any long-term grid down preparations. However, most large-scale solar backup power systems are expensive and will require a qualified dealer to install. In addition, EMP testing has shown many of the devices used in any solar power system contain microelectronic components which were destroyed, making the solar system unusable. This chapter will describe the function of each component of a solar backup power system, and ways to protect these solar components from an EMP.

SOLAR PANEL

A solar panel usually has an aluminum frame, a tempered glass front, covering a fixed number of individual solar cells and their interconnect wiring, which is enclosed on back with a vacuum-sealed vinyl sheet. The back may include a junction box to make electrical connections to the internal positive (+) and negative (-) terminals, or a sealed junction box with two molded interconnect cables having polarized male and female connectors.

Most lower voltage solar panels have an open circuit voltage (not connected to any load) of 21-volts DC. When connected to an external load, these solar panels will operate around 17 volts, which is ideal to charge a 12-volt battery. Most solar panels over 200 watts in capacity will normally have an open circuit voltage of 45 volts and

CHARGING BATTERIES WITH SOLAR POWER

will operate around 37 volts when connected to an external load which is too high to charge a 12-volt battery. These higher voltage panels are normally wired in series with other solar panels to produce over 400 volts DC to supply a high-voltage inverter selling power directly back to the grid and not charging batteries. The biggest mistake most solar novices make is trying to use these higher voltage solar panels designed for grid tie systems to charge 12-volt DC batteries.

CHARGE CONTROLLER

A solar charge controller is similar to the voltage regulator in a car that takes the constantly changing voltage and current output from the alternator as the engine changes RPM and stabilizes the charging voltage and current going into the battery. All solar power systems that charge a battery will require a solar charge controller. Lower cost charge controllers will not have any meter display and work basically like an on-off switch. Their only function is to automatically allow solar electricity to pass through to the battery when the battery voltage is lower than the solar panel voltage and turn off the charging when the battery voltage is high. This basic charge controller cannot adjust the voltage or current flow going into the battery from the solar panel, so these should only be used in very small solar applications.

The slightly more expensive pulse width modulation (PWM) charge controller is also basically an on-off switch wired between the output terminals of the solar panel and the battery. However, this charge controller monitors the battery voltage. The full output voltage from the solar panel going into the battery is constantly being turned on and off (pulsed) at a different rate and length (time) of pulse based on the charge level of the battery. This allows a gradual tapering off of the charging process as the battery approaches a fully charged state, which extends battery life and reduces battery water loss or out-gassing.

The maximum power point tracking (MPPT) charge controller is the most efficient solar charge controller currently available and includes multiple program functions. Since this charge controller can be programmed to adjust the battery charging process based on battery type, battery temperature, and battery state of charge, it will provide the fastest charging with the least risk of battery damage from over-charging or over-heating. It's also extremely expensive and normally used only with larger arrays and 24 or 48-volt DC battery banks.

SOLAR and DC FUSES

Any positive wire from an electrical device that is connected to the positive (+) terminal of any battery should include an in-line DC fuse. All 120-volt AC electricity consists of a voltage flow that "reverses" or alternates sixty times each second. This means multiple times each second the voltage is actually Ø for a fraction of a second. This also means it's fairly easy for any AC rated fuse or circuit breaker to extinguish the arc that develops as the contacts separate under an oversized current.

However, DC electricity being supplied from a solar panel, solar array, or battery does not alternate or cross Ø voltage at all and remains a constant current flow. Like trying to put your hand over the end of a high-pressure water spray, DC electricity is difficult to stop once it is flowing, especially if under an overload condition.

Do not use circuit breakers and fuses rated for 120-volt AC wiring to protect DC wiring loads. As the contacts start to open, it's not unusual for the heavy arc to "weld" the contacts together, or just blow right through an AC fuse. While automotive, RV, and boat fuses are DC rated and will safely protect DC loads if properly sized,

CHARGING BATTERIES WITH SOLAR POWER

it should be noted this type fuse is not approved by the NEC to protect permanent house wiring circuits.

SOLAR BATTERY

While battery technology has undergone some major advances to increase power density and reduce weight thanks to electric car research, battery weight and physical size are usually not as important in a residential off-grid solar and emergency backup power system. If you only need a limited capacity solar backup battery, the most practical battery to use is the 12-volt AGM Group 31 deep-cycle RV/boat battery. These weigh sixty-five pounds and are totally sealed so they will not spill acid or vent explosive gases if charged properly. These batteries are typically used to power 12-volt appliances in an RV or the trolling motor in a boat for hours before needing recharging.

The next most common deep-cycle battery used to power multiple loads is the 6-volt T-105 (golf cart) battery weighing sixty-three pounds. These batteries have very thick lead plates, which are designed for a daily deep discharge and recharge cycling, but most types have fill caps and not sealed like AGM or GEL batteries. Since this is a 6-volt battery, you will need two batteries that are wired in series to power any 12-volt DC loads, and these must be located in a vented room due to out gassing during the charging process. These batteries are fairly inexpensive and it's not unusual to find eight wired in series to power a larger 48-volt DC inverter having a 120-volt AC output capacity up to 4,000 watts.

Just before making a final connection to any battery post, be sure to clean both the battery post and the wire terminals with a stiff wire brush to a bright and clean condition to minimize resistance, as lead oxidizes very quickly. The lead battery post will be bright silver in color when cleaned properly.

CHARGING BATTERIES WITH SOLAR POWER

BATTERY INVERTER

The inverter converts a 12-, 24-, or 48-volt DC battery voltage, depending on the model, into a 120-volt, 60-cycle AC power. This output can power any lights and electronic devices you have that require 120-volt AC power. While inverters come in all sizes, most grid down applications may only require an inverter under a 300-watt output to power small 120-volt AC loads, such as a flat screen television, laptop computer, satellite receiver, or internet router. Inverters this small are normally designed to plug into a vehicle's 12-volt DC dashboard outlet, which is perfect for most mobile and bug-out power requirements.

Larger inverters will require a direct power connection to the vehicle's battery, and this wire can be fairly large if the inverter wattage exceeds 1,200 watts. Inverters 2,400 watts or larger need to be powered by a 24-volt or 48-volt battery since a 2,400-watt load at 120-volt AC translates to a 200-amp load on a 12-volt battery! Since all inverters have an efficiency loss when converting from DC to AC power, I would not use an inverter supplied from a 12-volt battery, to power a 120-volt AC charger, to charge a 5-volt DC cell phone! Use a 12-volt DC charger to directly charge a cell phone from the 12-volt dashboard outlet.

Any electric coffee pot, toaster, or hot plate, and any major household appliance including refrigerators, washing machines, and air-handling units would quickly discharge a single 12-volt battery. These large loads should only be operated when generator power is available during a grid down event, and never powered from an inverter and battery unless you have a large whole-house solar power system.

In addition to different wattage ratings, inverters are advertised as having either a modified or pure sine wave output. The pure sine wave inverter produces a 120-volt, 60-cycle AC output that is almost exactly the same power quality as grid power. Any television,

CHARGING BATTERIES WITH SOLAR POWER

electronic device, or battery charger will function normally on a sine wave inverter. A modified sine wave inverter requires fewer parts to manufacture and is much less expensive than a pure sine wave inverter.

The 120-volt AC output of a modified sine wave inverter is actually a series of voltage steps, which approximates the 60-cycle sine wave profile of the utility grid but not smoothly. Most lighting fixtures and motor-driven appliances will work normally on a modified sine wave inverter, but some more sensitive electronic devices only work on the more expensive pure sinewave inverters.

Fortunately, many of today's smaller electronic devices and laptop computers with rechargeable batteries have power supplies designed to work with either 120-volt U.S. or 240-volt European electrical systems. These power supplies are much more forgiving of power quality and voltage fluctuation and will usually operate just fine on either inverter type.

Inverters designed for mobile application do not normally make the best residential off grid inverters since they do not include a built-in battery charger or automatic transfer switch. Inverters that are designed for a residential application usually have a higher wattage capacity since they are not limited to a 12-volt vehicle battery voltage. Having a built-in battery charger allows the inverter to charge the battery directly when either grid power or generator power is available.

PORTABLE SOLAR

Most readers are only interested in keeping small electronic devices including a battery-powered radio, cell phone, and a few LED lights charged. This does not require a roof full of solar panels, a large wall-mounted inverter, and a thousand pounds of batteries typically found in an off-grid solar home. The easiest solution is having several sizes of foldup solar panels which can be unfolded, plugged into the device, and positioned to face the sun. A 10-watt solar panel will easily charge a cell phone, and a 25-watt panel will charge a laptop computer in one afternoon.

For larger electrical loads including medical devices, ice makers, and power tools as discussed briefly in prior chapters, I recommend a single rigid solar panel in the 125- to 150-watt capacity range, a 20-amp PWM solar charge controller, and a Group 31 deep-cycle AGM battery as previously described.

The charge controller, fuse, and battery can be mounted in a non-metallic battery box typically used for boat trolling motor batteries and connected to the separate solar panel with twenty or more feet of 14-gauge wire. I have much more wiring and construction details to assemble a portable solar power station in my book *Lights On*, and I recommend adding this book to your preparedness library.[1]

SOLAR SYSTEMS VERSES EMP

Tests have shown larger solar arrays, solar chargers, and power inverters designed to provide backup power for an entire house or power an off-grid residence will *not* survive an EMP or solar storm without significant wiring modifications. Above roof array wiring will require adding multiple ferrites on each solar panel and a metal oxide varistor (MOV) added to each wire entering and exiting the solar charge controller and inverter. A more robust system of ground rods and DC surge arrestors will also be needed.

If you are planning to install an inverter and solar array large enough to provide backup power for an entire house, Sol-Ark inverters have been tested to withstand an EMP or solar storm event without damage. This Texas company also offer kits to harden the

CHARGING BATTERIES WITH SOLAR POWER

solar array, solar charger, and battery bank from the high voltages and currents induced into the wiring by an EMP.

No matter which type and size solar backup power system you choose, it should be clear once the grid is down for months, regardless of the cause, and all generators are out of fuel, the only realistic way a homeowner can have any electrically powered appliances is with batteries and solar chargers. This means making sure your solar hardware is protected from EMP damage, regardless of system voltage or size.

CHAPTER 29

Which Rechargeable Batteries?

In prior chapters we have reviewed many types of DC-powered devices that can replace their 120-volt AC counterparts during an extended power outage. I have also pointed out how battery-powered devices must be more energy efficient than grid-powered devices to maximize battery life. This makes battery-powered devices a better choice to have during an extended power outage, than trying to keep grid-powered devices operating using an emergency generator or backup power system. Their smaller size and portability are also ideal if you do have to relocate or bug out during some type of disaster. Due to their small size, many of these electronic devices will survive an EMP, so which batteries should you standardize with to keep things operating smoothly if the grid is down?

Until now we have not actually discussed battery technology, or which types of batteries to look for when purchasing battery-powered devices. Of course, you could just buy a fresh pack of batteries every few months, just in case there will be a power outage in the near future. However, every time I have tried this approach, they either have lost their charge or started to corrode while still in the package by the time I needed them.

While it is still a good idea to keep a few packs of non-rechargeable AA and C size batteries around for your small flashlights and portable radios, converting over to rechargeable batteries is a

WHICH RECHARGEABLE BATTERIES?

better long-term solution. This is especially true for digital cameras and similar devices that seem to eat batteries. The first digital camera I owned ran on four AA size batteries and I could easily go through sixteen batteries in a day of project documentation. Cell phones and portable computer devices are now manufactured with longer life internal rechargeable batteries that stay in the device during charging and cannot be removed. However, most portable battery-powered flashlights, radios, walkie-talkies, digital cameras, and remote controls are normally furnished with replaceable AAA, AA, or C size disposable batteries.

Think of all the times you needed a flashlight, only to find it buried in a kitchen drawer with dead batteries. The simple act of buying a pack of flashlight batteries the next time you go to the store will not be possible during a grid down event. Did you ever go to a grocery store the day before a snowstorm was forecast? The first things to sell out will be bread, milk, bottled water, and flashlight batteries, so you can imagine how batteries will be impossible to find during an extended power outage that could last weeks or even months during a true grid down event. How will you power all those battery-powered devices once their disposable batteries die?

Discarded button batteries containing mercury-oxide contribute almost 90 percent of the mercury going into today's landfills, while dry cell batteries contribute half of all cadmium and nickel found in our landfills. It is estimated the United States discards over three-billion dry cell batteries each year, which is over thirty-five batteries per family.[1] While this may be a reasonable national average, I think for many homes with teenagers the number is far higher!

To extend the life of disposable batteries, many manufacturers are switching to heavy metals, which can include nickel, cadmium, lead, mercury, and acid, in their battery construction. If discarded in landfills, these potentially toxic materials can leach into our lakes or groundwater, or if incinerated can become harmful airborne ash.

WHICH RECHARGEABLE BATTERIES?

If you are like me, your first impression of rechargeable batteries was not good. Every time you needed to use a flashlight or digital camera having rechargeable batteries, the batteries were dead, and you had to wait all day for them to recharge. Today's rechargeable batteries are different, and you really need to start using them on a regular basis in all your battery-powered devices, as they will be priceless during a grid down event. To make this change, you need to first purchase a minimum of two sets of rechargeable batteries for *each* battery-powered device you have. Leave one set in the device, while the other set can be charged up and ready to switch out.

Second, you need one of the new multi-battery chargers which can charge from eight to twelve batteries at the same time. It makes no difference as to the mix of battery types or sizes. As soon as a battery is placed in these new chargers, they first run a diagnostic test to determine the type of rechargeable battery, its existing level of discharge, and then selects the best procedure to use for recharging each battery as fast as possible and without risking battery damage from over-heating. Some lower cost battery chargers do not monitor temperature during charging and can damage the battery due to improper charging voltage or when left too long in the charger.

You need a central location for your battery chargers and storage for any already charged batteries waiting to be switched with the discharged batteries just removed from a device. Since you may also have a separate cell phone charger, laptop computer charger, and perhaps a charger for a Bluetooth device, keep all of these chargers together. Use a switched outlet strip to power all of these chargers which allows turning everything off when all charging has been completed. These save wasted energy as most of these chargers still consume power even when not charging.

WHICH RECHARGEABLE BATTERIES?

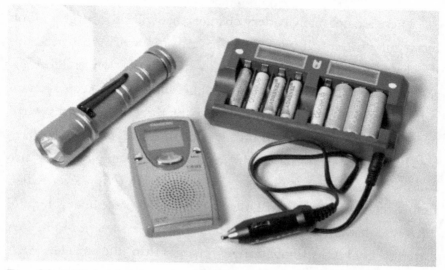

Fig. 29-1 Eight AA battery capacity charger and useful battery-powered devices.

Finally, you also need lots of rechargeable batteries. Many of us grew up with the older Ni-Cad battery technology which had "memory" problems, and if not fully discharged before recharging, the charging process would end before reaching a full charge. In addition, most of the earlier rechargeable battery technologies would lose the charge in a few days and needed to be recharged again immediately before using.

Today's rechargeable batteries are higher quality and have a much longer life than earlier designs. The most popular rechargeable batteries sold today are the Nickel-Metal-Hydride (Ni-MH). The Nickel-Cadmium (NiCd) rechargeable battery is an older technology and is being replaced by the newer Ni-MH batteries. The Ni-MH battery technology will store two to three times more energy than the older NiCd technology and can be recharged more than one thousand times. Most rechargeable Ni-MH batteries will hold their full charge for at least six months.

WHICH RECHARGEABLE BATTERIES?

There are not many battery chargers that will also charge a 9-volt rectangular battery. If you do not have many devices that need this size battery, it may be easier just to switch to a longer life non-rechargeable brand having a much longer life and just keep extra on hand. Try to standardize on only two or three battery sizes when purchasing any new electronic devices to make it easier to have freshly charged batteries ready for use. For example, AAA size batteries are very small and have limited operating time even when freshly recharged. Some LED flashlights and digital cameras require AAA size batteries, and some require AA size batteries. The AA batteries hold substantially more charge than the smaller AAA batteries, so the larger AA battery should be the minimum size of rechargeable batteries you use.

Larger LED lanterns and battery-powered radios typically use either C size or D size batteries. You will probably find more LED lanterns and portable radios using C size batteries these days than the larger D size to reduce device size and weight, so the C size rechargeable battery should be your second rechargeable battery size to have on hand.

When I started my conversion to all rechargeable batteries, I found a wide difference in charging capacity from one battery manufacturer to another. For example, the highest quality C cell rechargeable batteries from Sanyo have a 6,000 milliamp-hour (mAh) rating, while some less expensive rechargeable C-cell batteries from other suppliers are in the low 4,000 mAh range. While less expensive, this 33 percent lower charge capacity would translate into a similar drop in device operating hours.

Since these batteries have a much smaller charge capacity than the amp-hour ratings used for comparing larger vehicle batteries, the smaller scale milliamp-hour (mAh) rating is used, which is just amp-

WHICH RECHARGEABLE BATTERIES?

hours divided by 1,000. Any AA size battery with a 3,000 mAh rating will have twice the stored power of a similar-sized battery having a 1,500 mAh rating. Unfortunately, these ratings are not always easy to identify without reading the small print with a magnifying glass. When you are shopping for rechargeable batteries, remember, the brand that cost half as much may also have half the mAh recharge capacity.

Some rechargeable battery brands and models consistently outperform all others. Professional photographers needing reliable flash performance, first responders needing life-saving portable two-way radio communications, and field technicians needing reliable battery-powered test equipment soon learn which rechargeable batteries give the best service. Based on their recommendations and my own experience, I have found the Panasonic and Sanyo Enloop rechargeable batteries consistently rated highest for charge capacity and reliability.

There may be other brands equal or better than these, so check mAh charge capacity and customer ratings before making a purchase. If you are testing batteries with a voltmeter or battery tester, keep in mind all AA, AA and C size Ni-MH batteries will actually measure 1.2 volts, not 1.5 volts as typical of non-rechargeable batteries when fully charged.

For the ultimate in being prepared, some models of multi-battery chargers include both a 120-volt AC power adapter and a DC adaptor to fit the 12-volt utility outlet in your car or truck. Since the 12-volt DC adapter will easily mate with the power connection on a foldout solar charger, this will allow keeping all of the smaller batteries for your flashlights and small electronic devices charged over and over again using grid power, generator power, your vehicle, or the sun. This is one grid down preparation you can do now, which

WHICH RECHARGEABLE BATTERIES?

will provide immediate savings in battery replacement costs while getting you ready for a time when disposable batteries are no longer available at any price.

The internet and magazines are full of ads for solar generators and portable backup power units, typically the size of a desktop tower computer. These can be convenient for a power outage lasting a day or two, especially if you do not own a generator. They also are ideal for someone with limited storage space or are less experienced with wiring as these units are self-contained. Everything is prewired and the single case includes a battery, 120-volt AC inverter, and battery charger to recharge from the grid when available.

Most models also offer an optional solar charger and separate solar module with enough cable to locate outside while the unit stays indoors. My biggest problem with these units is their battery capacity and product claims. Their advertising list all of the household appliances they can power, including a coffee pot and a refrigerator! While this is technically correct, the small print will usually show the actual run time for anything larger than a flat-screen television will probably be a few hours, not days or weeks. For the most benefit, only use these self-contained solar battery units to power smaller electronic devices and be sure to have a solar panel large enough to fully recharge in a day.

The lithium-ion (Li-Ion) batteries typically found in today's cell phones, laptop computers, and portable power tools provide even more charge capacity, are smaller, and weigh much less. However, most of these are either built into the device or are supplied with their own special charger that matches their unique shape and charging cycle requirements.

As I discussed in chapter 11, the newest battery technology being used to power all electric vehicles are the lithium-ion cells which are

WHICH RECHARGEABLE BATTERIES?

slightly larger than a AA flashlight battery, with up to five hundred wired together to make up a single battery tray. Lithium-ion batteries provide the highest energy density available today, which is why this is also the most popular battery used to power cell phones. Unfortunately, lithium-ion batters do not like high temperatures and have been known to catch fire if charged while very hot or very cold. As discussed in chapter 11, the multiple battery trays in any electric vehicle each require a battery management system (BMS) to provide much more control over the charging and discharging process and battery temperature.

Manufacturers of these vehicle tray batteries are starting to offer a wall-mounted version for emergency backup and solar power systems. These do allow a much heavier discharge each cycle, and far more charge/discharge cycling than more traditional lead acid storage batteries. Unfortunately, they are still extremely expensive and their EMP survivability is a concern due to the need for an electronic battery management system (BMS).

To offer more battery stability, the lithium iron phosphate (LiFePO4) battery technology was developed. While this battery has a third less energy density than the lithium-ion battery, they can provide many more deep-discharge cycles and are much less sensitive to temperature extremes.

I mention this battery because some manufacturers are assembling the small individual cells into a battery case that is almost identical in size and shape as a standard 12-volt deep cycle RV/boat battery. Unlike a convential AGM or GEL battery which will have a significantly shorter life if repeatably discharged below 50 percent, lithium iron phosphate batteries can be repeatably discharged up to 90 percent and still provide several thousand charge-discharge cycles. Unfortunately, like the lithium-ion battery, these newer batteries still

WHICH RECHARGEABLE BATTERIES?

require a battery management system (BMS) which are built into the battery case.

The 12-volt lithium iron phosphate battery is a simple one-for-one replacement for a group 24 or 31 size RV/boat battery and will provide an instant improvement in depth of discharge and cycle life, but they are still almost five times the cost. Many RV owners having a roof-mounted solar array are starting to convert over to these lithium iron phosphate 12-volt batteries as they provide far more amp-hour storage with half the weight.

There is the survivability issue with these lithium-based batteries since each require a built-in BMS consisting of microelectronic components which are highly susceptible to EMP damage. While a disconnected 12-volt lithium iron phosphate battery is too small to couple with an EMP voltage spike, their BMS will be susceptible to EMP damage if connected to multiple long wires or chargers when an EMP strikes.

Any rechargeable battery, regardless of its chemistry, is basically operating like the gas tank in a car. It holds a fixed number of "amp-hours" of stored energy just like the car's gas tank holds a fixed number of "gallons" of gas. Once all the stored energy is consumed, that's the end of any useful work until the battery can be recharged or the gas tank can be refilled.

While any rechargeable battery can be recharged, what comes out must be replaced, which means however many amp hours were supplied to power electrical loads must go back in, with an equal number of amp hours of charging plus extra to account for up to a 20 percent combined efficiency loss in the charging process, which includes converting electrical energy into chemical energy, then chemical energy back into electrical energy to power a device.

WHICH RECHARGEABLE BATTERIES?

While some of the newer battery technologies can be recharged to a full state in only a few hours, most conventional deep-cycle batteries can take up to eight hours to be fully recharged. If you could discharge a battery in a backup power system in only an hour or two when powering a large appliance, it could take more than a day to fully recharge the battery again depending on the size and battery technology. This is especially true when trying to recharge a battery with a solar panel and only four or five hours of solar availability each day, depending on your location and weather. Regardless, if wanting to purchase a pre-assembled solar battery or assemble your own from a deep-cycle battery and separate inverter and solar charger, everything will depend on how big the gas tank is (amp-hour rating of the batteries) and what you will have to refill it (wattage output of the solar panel).

CHAPTER 30

Connecting Battery-Powered Devices

When you want to power a conventional television, kitchen appliance, or hairdryer, you just plug them into the nearest 120-volt AC wall outlet. You don't need to think how the electricity got there, where it was generated, or if there will be enough additional capacity to power the added device. It just works, at least until it doesn't!

Most European countries have standardized on a nominal 240-volt, 50-cycle AC electrical power using a two-prong, un-grounded wall receptacle, although the actual spacing of the two-prongs in the plugs will vary from country to country. The United States standardized on a nominal 120-volt, 60-cycle AC electrical distribution system from the very beginning, although after the 1950s a third pin was added to all receptacles and smaller wattage appliance plugs for an earth ground. Unlike Europe, all of the wall outlets in the United States use a standard spacing of the outlet and plug blades, although the neutral blade is slightly wider, ensuring a two-pin plug will be oriented correctly so the neutral blade is always connected to the neutral conductor.

Most small household appliances have housings made from non-conductive plastic that provides a second layer of electrical insulation, which allows using the two-blade polarized plug. Larger appliances having a metal housing require a three-bladed plug. The added ground pin bonds all non-energized metal parts together to reduce

CONNECTING BATTERY-POWERED DEVICES

the risk of shock. All two-bladed and three-bladed appliance cords will fit into the same 120-volt AC wall outlets.

Battery-powered devices are different, as most operate from internal batteries and are only plugged into a power source to recharge. Most battery-powered devices will operate on external grid power when their chargers are left connected to a wall outlet, which avoids discharging their internal battery during normal operation. To reduce the size of a portable device and accommodate the different grid voltage standards, chargers for these battery-powered devices are typically part of the external power cord or part of the end plug.

This can be a nuisance since you will need to have a separate charger and dangling power cord for each battery-powered device you own, which needs to be found and connected whenever the device has to be recharged. Having a separate external charger and cable for each device does allow using multiple sources of power to recharge the same device as in-line chargers are available to match almost any power source, including the 12-volt DC dashboard outlet in your car. Unlike most 120-volt AC appliances, being able to remove the charging cable from any battery-powered electrical device also reduces its vulnerability to EMP damage.

One thing I noticed when I first started acquiring more battery-powered devices was the large collection of battery chargers I had accumulated, which are all black and look almost identical. Unfortunately, since they may all look alike, they each could have a different style plug, different charging voltage, and sometimes a very different current rating. To simplify finding the one I need, I used a silver color permanent ink marker to write the device name on the side of its matching charger.

While AC appliances will only operate when connected to a 120-volt AC power source, any battery-powered device can be recharged from the 12-volt DC utility outlet in a car, truck, boat, RV, a solar panel, as well as a 120-volt AC grid or generator power

CONNECTING BATTERY-POWERED DEVICES

source when available, as long as you have the correct charger and cable. The downside is you will need a different charging device for each source of power, and there are many different types of plugs used to connect a charger to each battery-powered device. This lack of standardization makes it harder to use the same charger for multiple DC powered devices.

During a true grid down event, flexibility is very important and its likely you will be called upon to do lots of improvising to keep all battery-powered devices operational. To make this easier, it's important to select those electronic devices that allow standardizing on the same type connectors and batteries so they can share the same battery chargers. Also remember, wire size is determined by the amp flow it must carry not the voltage. For a given watt load, it takes ten times more amps at 12 volts than at 120 volts, so DC wiring will be a larger gauge to supply the same wattage load. Most of the rechargeable LED lanterns, portable radios, and smaller electronic devices that use rechargeable batteries have a barrel connector to connect an external charger.

Unfortunately, these are manufactured in multiple diameters, different lengths, and even different sizes of the internal pin. Most connectors are polarized with the center pin as positive and the outer shell as negative but verify before using a different charger or charging cable.

The 12-volt DC plug for charging all battery-powered devices in a vehicle will be a "cigar" plug, which fits the dashboard

Fig. 30-1. 12-volt DC barrel plugs are typically found on smaller battery-powered radios and rechargeable LED lanterns.

238

CONNECTING BATTERY-POWERED DEVICES

outlet in your car or truck. Every car or truck built since the Model T includes a dashboard "cigarette lighter," which is why today's vehicle manufacturers still include the 12-volt DC socket as standard to power 12-volt DC accessories, but they no longer provide the lighter! No doubt you have a 12-volt DC charger for a cell phone or GPS receiver plugged into your car's dashboard outlet right now.

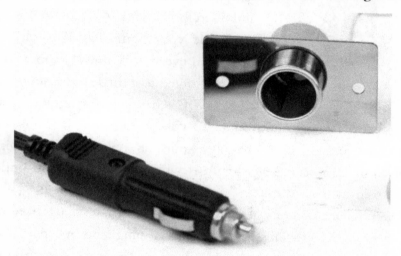

Fig. 30-2. 12-volt DC "cigar" plug is the standard dashboard outlet for all vehicles.

My truck came with three different 12-volt utility outlets, and my car has a 12-volt utility outlet in both the front and back seats, plus an additional outlet inside the center console. Boats, recreational vehicles, ATVs, and truck campers also have one or more 12-volt DC dashboard outlets.

Since you will have multiple battery-powered devices that will need to be charged at the same time during a real emergency, I recommend keeping several multi-outlet adaptors in each vehicle. Most cell phones, rechargeable flashlights, and portable radios rarely draw more than one or two amps at 12-volts DC when charging. This means charging multiple devices from the same 12-volt DC

CONNECTING BATTERY-POWERED DEVICES

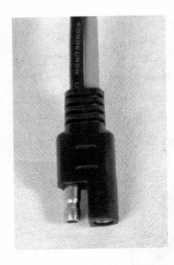

Fig. 30-3. 12-volt "SAE" automotive style polarized connector is now common on many brands of larger foldout solar panels and vehicle accessories.

dashboard outlet should not be a problem as most vehicle outlets are typically rated and fused for 15 amps.

While your home has 120-volt AC outlets on every wall of every room, you may be limited to a single point source of 12-volt DC charging power during an emergency. Having multi-outlet 12-volt DC adaptors will also save generator fuel or solar charging hours, as opposed to charging everything one at a time, which is why you need a multi-outlet adaptor in each vehicle.

Some manufacturers of foldup solar panels have standardized on the SAE quick-disconnect 12-volt DC plug and receptacle to connect their solar panels to other devices. These are fairly easy to recognize as they are flat molded plastic with the positive and negative pins side by side, but one pin is exposed. To avoid a mismatch of battery polarity, they cannot be plugged together with the pins reversed.

With the tremendous increase in the use of portable laptop computers, iPads, think pads, and smart phones, most

Fig. 30-4. USB type 5-volt connector is found almost exclusively on cell phones, portable computing devices, and small foldout solar panels.

240

CONNECTING BATTERY-POWERED DEVICES

manufacturers now furnish a USB connector on the end of the cable between the battery-powered device and the source of charging power. All USB outlets are 5-volt DC, not 12-volt DC. Any adaptor you use to connect a USB cable to a 12-volt DC socket must include a DC-to-DC voltage converter built into the plug or use an in-line charger to match the different DC battery voltages.

Larger 12-volt DC inverters require far more power to operate then can safely be supplied by a 12-volt DC dashboard outlet. These loads should be wired directly to the vehicle's battery using bolt-on battery cables. These cables should be as short as possible.

To address the problems of interconnecting more power demanding radios and transmitters to multiple sources of emergency backup power, the amateur radio community has standardized on the two-conductor red and black Powerpole in-line connectors which are available in higher 15, 30, 45 and even 75-amp capacities using polarized 12-volt DC red (positive) and black (negative) contact blades.

Alligator clips used to connect a device to the vehicles' battery should be considered a temporary battery connection and not used for continuous operation. When they do not make a good connection, they will overheat and can even melt the insulation on the wires.

Most vehicle batteries are either difficult to access in smaller car models or may have terminal corrosion which can make it difficult to achieve a good

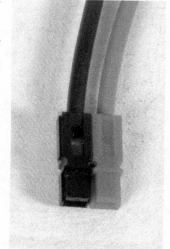

Fig. 30-5. Powerpole polarized battery connectors are available with ratings up to 75-amps and are often found on larger power demanding inverters and have radio equipment.

CONNECTING BATTERY-POWERED DEVICES

electrical connection. In addition, this type of temporary battery connection will not allow closing the hood, and the battery cables could tangle with the cooling fan as the engine will most likely be running when a larger capacity electric load is connected to the battery with alligator clips.

While the *National Electrical Code* specifies black for all ungrounded power conductors, white for neutral conductors, and green or bare wire conductors for ground in residential house wiring, most vehicle, RV, and boat wiring still follow the older standard of red for positive and black for negative. Although any color wire will still work, following these color standards will provide much safer future system repairs, especially when it's not clear which wire goes to a terminal that is not readily accessible.

CHAPTER 31

Protecting Your Vehicles from EMP

In chapter 11 I provided an introduction to the EMP risks to a modern car or truck, including all-electric vehicles. As noted, a modern vehicle is full of microprocessors, electronic sensors, and digital dashboard displays. EMP testing has been inconclusive in identifying which vehicles will suffer catastrophic engine failure, which will suffer minor loss of accessories, and which may not be affected at all.

Part of this uncertainty has been due to the much better radio frequency interference (RFI) shielding that has been recently added by manufacturers to their vehicle's wiring. There is concern a stray radio transmission or lightning strike could cause electronic braking systems to lock up, accelerator pedals to stick, or perhaps the engine just quit while going seventy miles per hour.

Manufacturers are relying more and more on computer controls to take over many functions previously operated manually by the driver. Since increased RFI shielding has occurred at the same time more computerized devices are being added, this could balance out with no net increase in EMP vulnerability. However, it is also possible the improved RFI shielding will not be enough if overwhelmed by the massive energy shockwave of a real EMP event. Vehicle manufacturers do have concerns regarding the RFI shielding in new vehicle designs. The results of new vehicle RFI interference testing are strictly guarded and not available to the public. It's

PROTECTING YOUR VEHICLES FROM EMP

common to see totally camouflaged new models on the back of a truck heading into RFI and EMP testing centers.

This leaves us with the uncertainty of how best to reduce the risk to our own car or truck. My 1970s soft top CJ-5 Jeep contains no electronics at all. It has a mechanical-driven ignition distributor and non-electric dashboard gauges and toggle switches. In fact, when using a car wash, I can hose down the entire interior, seats, and dashboard without causing any component damage or electrical shorts.

Other than the possibility of an EMP event blowing out the fuse to my headlights, it's an almost certainty that this old Jeep will still be running long after a grid down event, assuming I can still find gasoline! However, where does this leave the general public? After all, everyone can't go back to driving 1960s- and 1970s-era vehicles, let alone find ethanol-free gasoline, tires, and repair parts to keep them operating.

This is where a little mechanical ability can keep your vehicle on the road. The vehicle testing that was carried out by the congressional EMP Commission found that while all car and truck engines stopped when exposed to high levels of EMP energy, many could be made operational again by disconnecting the battery for a few minutes which caused all internal computer devices to reset.

Obviously, this will not repair any micro-electronic components that were actually fried by a high voltage surge in the wiring. If you do not know where your vehicle's battery is located, let alone have the tools to remove the battery connections and clean the terminals, this might be a good time to learn!

Although a modern vehicle will have multiple computerized devices and sensors scattered around the engine compartment or hidden behind the interior dashboard, there is always a main computer module somewhere that controls the engine ignition system, plus manages the other computerized control devices. While

PROTECTING YOUR VEHICLES FROM EMP

it would be very expensive to keep a spare for each of these separate computer control modules and sensors, at least having a spare ignition control module stored in a Faraday cage as discussed in chapter 36 might be a reasonable EMP damage solution.

Since these control modules are the size of a paperback novel and utilize a single snap on wiring harness, they are easy to replace. However, actually finding where these are mounted is not easy, although they are usually located somewhere inside the vehicle to protect it from the weather. I finally found the computer module for my diesel truck hidden behind a removable panel in the foot well of the passenger seating area. These units are extremely expensive if ordered from a local dealership but are usually easy to find and much less expensive from any vehicle salvage yard.

There are a few commonsense ways to reduce the impact of an EMP on your personal vehicle, and most do not cost anything. Whatever you can do to attenuate the strength of an electromagnetic shockwave before it reaches the microelectronics inside your vehicle will reduce the potential damage. For example, while more and more vehicle bodies are now made of plastics and fiberglass, selecting a vehicle that still relies heavily on all-metal body components, at least surrounding the engine compartment, will add EMP protection.

Small ferrites that will be discussed in far more detail in chapter 33 can be easily placed on each wire entering this ignition control module. While not guaranteed to make a vehicle EMP proof, anything that helps attenuate the initial E1 pulse entering into any electronic device will help.

Keep in mind a vehicle electronic control module with its metal enclosure is too small to be damaged by an EMP if it was just sitting on a work bench. But once installed, many of the connecting wires will be more than long enough to be energized by the E1 phase of an EMP and send this induced high-voltage into the control module.

PROTECTING YOUR VEHICLES FROM EMP

I have an all-metal prefabricated shop building with a high ceiling and only one window. I have found my cell phone is totally useless when I am working inside even though it's a fairly large space. While the electromagnetic energy from an EMP includes far more frequencies and at a much higher energy level than used for cellular communication, this still shows parking a vehicle in an all-metal garage does offer some EMP signal attenuation.

There are large covers now available that have good electromagnetic blocking characteristics which are large enough to cover a car or truck. When completely covering a vehicle and extending to the ground on all sides, these covers can provide an additional level of EMP shielding. To be the most practical, this cover should be used for a vehicle that is only driven occasionally and is kept parked in a garage when not being driven which will avoid storm or wind damage to the cover.

Finally, there is the simple fact that the orientation of the vehicle in reference to the source of an EMP, the "shading" effects of any large nearby structures or ground terrain, and the distance between your vehicle and the EMP source will all have some effect on the level of electromagnetic energy that reaches your vehicle. You obviously cannot leave your car locked up day and night in a metal building or select your parking space based on your guess as to the direction to a possible EMP detonation! This means there will always be a "luck of the draw" issue that some vehicles will suffer little or no damage from an EMP just by pure chance and others may never run again.

CHAPTER 32

Protecting Antennas and Shortwave Radios from EMP

In earlier chapters I discussed how an EMP or solar storm can damage electrical appliances and electronic devices by high-voltage surges coming in on the power lines and household wiring then into the electrical devices. I have also discussed how the much faster and higher frequency E1 energy can enter into electrical devices without traveling through the power cord. Any long wire or metal object can act like an antenna to absorb this electromagnetic shockwave and induce high voltages and currents into the electrical devices, even when they are disconnected from a wall outlet!

With the exception of a wired cable television antenna or satellite modem, most readers will not have any other electronic devices connected to an actual outdoor antenna. However, almost all shortwave radio listeners, and those operating CB base stations and ham radio transmitters will definitely have a large antenna either on a tall pole in the yard or on their roof. Hopefully the feed line from any outside antennas already have a basic lightning arrestor, but this will provide little or no protection from the high voltages induced into these antenna systems from the high frequency E1 phase of an EMP.

While most ham radio operators have already installed lightning arrestors on their antennas, and a good earth ground on the electrical wires powering their equipment, this will not protect against the much higher E1-induced currents and voltages coming in on the

PROTECTING ANTENNAS AND SHORT WAVE RADIOS

antenna cable, and E1- and E2-induced voltage surges coming in on the house wiring. Tests have shown a typical 50-ohm coax cable or transmission line will arc between the inner antenna wire and the outer shield when exposed to any high-voltage surge above a few thousand volts.

This means the voltage surge induced into an antenna from the E1 portion of an EMP event will most likely damage this feedline, which will act as a fuse to short out a major portion of the high-voltage surge before it reaches the more sensitive radio equipment.[1] In this case, using a less expensive coax cable will actually provide better protection from an EMP than a more expensive coax cable, but will require keeping a spare cable for replacement after an EMP event.

Obviously, radio operators should not rely solely on the internal arcing in a feedline as the only means of EMP-induced surge protection! The most effective protection against a high-voltage surge induced into the antenna system by E1 is to just not have the antenna connected to any radio equipment when not actually in use. This is much easier to accomplish by using an antenna selector switch

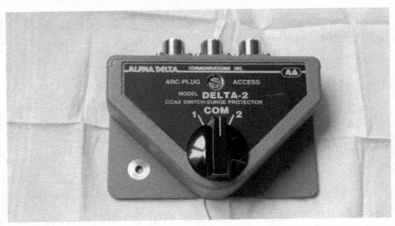

Fig. 32-1. Antenna grounding switch with built-in gas-discharge surge arrestor.

PROTECTING ANTENNAS AND SHORT WAVE RADIOS

that shorts to ground the feedline when switched off from the radio equipment. The second-best method is adding a gas discharge surge arrestor at the end of the coax antenna cable just before the connection to the radio equipment.

These devices are low cost and come with various clamping voltage ratings, which is the voltage the device shunts to ground any higher voltage trying to pass through in excess of this rating. Since a typical ham radio transmitter can send a frequency-induced AC voltage out to the antenna in excesses of 100-volts, the selection of a gas-discharge surge arrestor must have a voltage clamping rating above the output voltage of the

Fig. 32-2. Tiny gas-discharge surge arrestor the size of a peanut can shunt over 20,000-amps to ground!

transmitter. By far the simplest solution is to use an antenna selector switch that includes a built-in gas discharge surge arrestor like the Alpha-Delta antenna switch shown in the photo.

When in the off position, the radio equipment is disconnected from the antenna feed line, which is shorted to ground, but the gas-discharge surge arrestor remains in the antenna connection to the radio when in operation. A tiny gas-discharge surge arrestor about the size of a peanut can shunt to ground over 20,000 amps of current surge! However, it cannot turn on fast enough to respond to the fast acting E1-induced voltage which passes in a few nanoseconds. While this is true, tests have shown this method still works because the device is only part of the overall combined EMP protection.

For example, the insulation breakdown in the antenna feedline will block out everything above a few thousand volts. While E1

PROTECTING ANTENNAS AND SHORT WAVE RADIOS

energy does include a wide range of higher frequencies from 10 kHz up to 1 GHz, its strongest level is below 10 mHz. Most high-end ham radios include internal tuning circuits and filters that will further block out all frequencies outside the radio's normal tuning range. In other words, acting together, there is a fairly good chance the above steps will protect any CB base station, ham radio transceiver, or even just a shortwave listening receiver. For experienced ham radio operators, a four-part multi-issue article by past ARRL director, Dennis Bodson, published in the American Radio League's (ARRL) magazine dated November 1986 provides extensive technical instructions on protecting ham radio stations from all forms of EMP damage. I refer you to those articles for more in-depth information.[2]

Gas-discharge surge arrestors will need a clamping voltage above 300 volts to avoid shorting out the transmitter also connected to the same antenna, which can exceed 100-volts output when transmitting. However, ham radios and receivers containing microprocessors and computer components can be easily fried when exposed to much lower E1-induced voltages. In addition, external cables to code keys, headphones, speakers, and laptop computers can also serve as "antennas" to collect E1 energy and send into the radio equipment even if the outdoor antennas contain extensive surge protection.

For those of you with more electronic experience, by far the cheapest and easiest way to provide another layer of EMP protection is to install a metal oxide varistor (MOV) on each wire entering into the radio equipment. These are available with a wide section of clamping voltages that are far lower than a typical gas-discharge surge arrestor and will shunt to ground many thousands of amps in less than 20 microseconds. Back-to-back shunting diodes can also be used to protect smaller shortwave receivers, but these must not be connected to any transmitter output.

PROTECTING ANTENNAS AND SHORT WAVE RADIOS

Experience has shown metal oxide varistors (MOV) are one of best ways to reduce an EMP-induced voltage spike reaching and damaging internal microelectronic components. However, these EMP voltage spikes are essentially a DC voltage waveform, while MOVs are typically sized based on their response to an AC waveform. This can cause the MOV to heat up when blocking a large DC peak voltage.

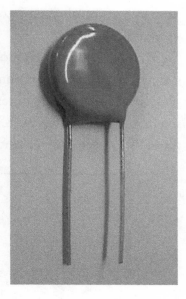

Fig. 32-3. Metal Oxide Varistor (MOV) the size of a quarter is available to clamp down at almost any desired voltage peak.

It is recommended to select an MOV having both a high temperature and high current rating. For typical 120/240-volt power wiring, a safety factor of three should be used when sizing an MOV, which gives a clamping voltage of 720-volts. Since most commercial electrical equipment must pass a 1,000-volt immunity test on their power input line, this equipment should be safe even above the 720-volt MOV clamping voltage.[3]

As noted in earlier chapters, E1 energy does not "couple" very well with small handheld cell phones, CB radios, shortwave receivers, FRS walkie-talkies, and 2-meter ham radios, if they are not connected to the antennas or power wiring. There is minimum risk of EMP damage to higher frequency walkie-talkies which typically have built-in antennas only a few inches long but would be a concern for a typical handheld CB walkie-talkie if the almost three-foot-long collapsible antenna is left extended. This also assumes these

PROTECTING ANTENNAS AND SHORT WAVE RADIOS

devices are not plugged into an external earphone or charger cable several feet long.

For maximum EMP protection of ham radio and large antenna systems, make sure everything is bonded to the same common earth ground. This includes antenna masts, antenna rotator, all radios, shields on coaxial feeds, telephone system grounds, satellite dish grounds, and house circuit breaker panel earth ground. Where

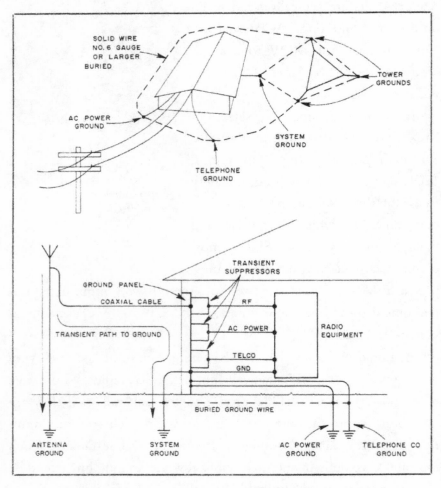

Fig. 32-4. Large outdoor antennas typically used for ham and CB radio hobbyists require an extra level of grounding to reduce potential EMP damage.

PROTECTING ANTENNAS AND SHORT WAVE RADIOS

multiple earth ground rods are used, tie all together with #6 AWG solid copper wire run underground to each other with no splices except at actual point of connection. Refer to the antenna grounding photo.[4]

While this chapter needed to be more technical to address ways to protect shortwave radios and outdoor antennas, it still may not provide a ham radio enthusiast with everything they need. I refer those readers to the Prepperdoc and Bodson articles referred in the footnotes for this chapter.

CHAPTER 33

Protecting Home Appliances from EMP

Probably the most reliable way to reduce the risk of damage to any electrical appliance from grid related brownouts, lightning spikes, and an EMP or solar storm is simply to keep them unplugged! Unfortunately, while this may be easy to do for a countertop coffee maker, it's not practical for larger appliances like televisions, microwave ovens, dishwashers, and clothes dryers.

When trying to protect the electrical appliances in a home, there are ways to reduce the damage caused by an EMP, but you must keep in mind there are three phases of an EMP shockwave (E1, E2, E3), and there is no single "magic black box" that will block all three. This means any total appliance protection solution will require more than one type of device or wiring modification. Tests have shown the maximum transient current any house wiring can carry from the main house panel to the electric outlets is 120 amps. While still high and no doubt will cause failure of some wiring, it's still not the thousands of amps proclaimed by some texts.[1]

My first recommendation to reduce EMP risk is to install a quality whole-house surge suppressor. However, before you head to Amazon, a word of warning. Being able to shunt to ground large voltage and current surges requires a device having robust and long-life internal components. Most whole-house suppressors from major name-brand manufacturers currently cost between one hundred and three hundred dollars. These higher cost devices include high-

PROTECTING HOME APPLIANCES FROM EMP

temperature rated metal oxide varistors (MOV). Lower cost MOVs can actually overheat and even self-destruct when subjected to very high voltages.

I think you will find any of the following whole-house surge suppressors to be reasonably priced and backed by well-established manufacturers:

Eaton # CMSPTULTRA

Siemens # FS140

Square D # SDSB80711

Intermatic # IG1240RC3

Those devices advertised at a much lower price may not be as reliable or offer the same level of protection, so check product specifications and buyer reviews before ordering. I have also noticed several new manufacturers offering an EMP "magic box" typically selling for hundreds of dollars more that do not appear to offer any additional EMP protection than you would get from a surge suppressor from more established manufacturers. In addition, marketing claims that they meet or exceed "all government EMP protection requirements" can be somewhat misleading so you need to read the small print.

Government EMP testing specifications referenced by some EMP protection manufacturers actually relate to hardening military communication systems and were never intended to test household wiring. In addition, these tests relate to blocking E2- and E3-induced voltages coming in on a power line and do not address high-frequency voltage surges from E1 energy traveling through the air and not through the house wiring!

I remember as a kid going to the state fair each fall when the leaves start to turn and the nights were getting cold. This was a big deal, and I couldn't wait to eat a candy apple and walk the long midway which had all kinds of exciting displays and booths offering the newest gadgets and games of chance. I remember there was

PROTECTING HOME APPLIANCES FROM EMP

always this fast-talking guy along the midway standing next to his car with the big V-8 engine.

The hood would be up, the engine was running, and hanging down from the hood was a huge tachometer showing the engine's RPMs. As soon as he had attracted a small crowd, he would take this little device that looked like a small juice can, slip it quickly between a spark plug wire and the distributor, and magically the engine's RPMs went up.

Of course, you can do the same thing by pressing down on the accelerator pedal, but never mind, this little device would increase engine power and double your gas mileage! Most likely these things died a few months later and stranded you in the middle of nowhere, but by then he is at another fair many states away. There will always be somebody hyping a product that is just too good to be true, and usually the manufacturer will be out of business in a year so forget about guarantees and warrantees.

Whole-house surge suppressors are usually mounted inside or attached directly on the side of your home's main circuit breaker panel and wired to a dedicated two-pole circuit breaker. It's important to remember there are exposed bare terminals and metal buss bars inside all breaker panels that stay energized even when the main circuit breaker is switched off. If you are not trained to work around energized electrical systems, I strongly suggest hiring an electrician to install this device, which in most cases will take them less than thirty minutes.

As noted at the start of this chapter, EMP is not a single thing, so a whole-house surge suppressor is not capable of switching fast enough to block the much higher frequency-induced voltages and currents induced into the appliance wiring by the early E1 phase of an EMP.

PROTECTING HOME APPLIANCES FROM EMP

Fig. 33-1. *Whole-house surge suppressor installed at main circuit breaker panel will reduce high-voltage surges from E2 and E3.*

Fortunately, there is a simple device that can dampen those high-frequency-induced voltages and currents which are very low cost and simply snap around the insulated wires to be protected. These are called ferrite chokes, ferrite beads, or simply ferrites, and they look like a round tube or doughnut that has been cut in half with a hinge joint on one side and a snap closure on the other side. These consist of a ceramic and iron compound and require no actual electrical contact to the wires being protected.

Since these ferrites are not actually electrically connected to the insulated cables they surround, it is sometimes hard to envision how these little objects can possibly provide any EMP

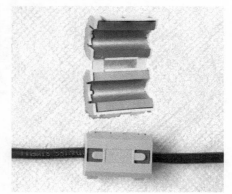

Fig. 33-2. *Ferrites around insulated wires will attenuate higher frequency E1 voltage surges.*

257

PROTECTING HOME APPLIANCES FROM EMP

protection to the equipment wired to the cables, but they do. While a quality surge arrestor can provide reasonable protection to these cables from the lower frequency E2- and E3-induced voltages and currents, they can do nothing to block the very high-frequency E1-induced voltages or currents. Remember, an EMP shockwave contains a wide range of frequencies and different peak rise times.

Think of the high-frequency-induced E1 wave "riding along" with the separate lower frequency power wave flowing past the small ferrite. As the two combined waveforms try to pass through the ferrite together, the high frequency waveform "energizes" the ferrite which basically absorbs this energy passing through the wire while the lower frequency electrical power continues on unaffected.

Ferrites are designed with different frequency induction characteristics, but if subjected to extremely high levels of RFI or EMP energy, they can overheat and become saturated, which drops their efficiency by up to 90 percent. Most ferrites intended for power wiring are formulated to have their highest induction or attenuation in the upper 1 MHz to 1 GHz range. This is at the upper range of frequencies generated by an EMP event.

The wire they protect are insulated, so the ferrite just fits around the insulation and snaps closed. These come in multiple sizes to fit loosely around any wire and are available in a broadband version and a high-saturation version. They typically cost under ten dollars. Make sure you order ferrites with the high saturation ratings.

Fig. 33-3. High-saturation type EMP blocking ferrite

PROTECTING HOME APPLIANCES FROM EMP

The smaller size ferrites are normally sold to block electrical "noise" or radio frequency interference (RFI) entering a radio from the power chord.

You will need a separate ferrite around each of the L1, L2, N lines supplying your home's main circuit breaker panel from the meter base, plus the ground wire to block E1 energy from the grid. Remember, you are working inside an electrically energized panel, so you might want your electrician to install these ferrites at the same time they are installing the whole-house surge suppressor.[2]

Adding a surge suppressor and ferrites at the main circuit breaker panel in your home addresses the high currents and voltages entering from the power lines and exterior wiring, but what about the downstream wiring inside your home? There are hundreds of feet of interior house wiring between this main breaker panel and all of the wall outlets and light fixtures scattered throughout the house.

There are also the additional power cords attached to each appliance and all audio and video equipment, and this equipment contain lots of microelectronic components. House wiring makes an excellent "antenna" to collect the high-frequency E1 energy passing through the air and into this wiring to reach these devices without needing the home's grid connection. A whole-house surge suppressor cannot block E1 energy entering this interior house wiring.

There are several additional steps you can take to protect more sensitive electronic devices, but these require installing electrical components inside the electric device. Adding a shoebox size computer UPS unit to supply more sensitive audio and video equipment will also serve to dampen high-voltage surges from an EMP, in addition to keeping the systems operating through a brief power interruption.

The metal oxide varistors (MOV) and gas discharge surge arrestors described in chapter 32 can be attached to each power wire

PROTECTING HOME APPLIANCES FROM EMP

entering any electronic device and shunt to earth ground any voltages and currents higher than the electronic circuits were designed to safely handle. These cost less than five dollars each and are able to handle these high currents because the peak duration of any EMP-induced voltage surge typically lasts only a tiny fraction of a second.

While these inexpensive shunting devices can be used to protect almost any electrical or electronic device from the damaging effects of an EMP, their actual sizing and specifications are based on many variables including frequency, rise time, cutoff voltage, and other design variables, which are beyond the scope of this book. I do not recommend that you install these blocking devices on any wiring yourself unless you have a strong electronics or ham radio background.

There will still be a high risk of E1 destroying microelectronic components, especially if the devices are connected to external power wires or an antenna. For total protection, you should consider storing a spare device in an EMP-proof container as discussed in chapter 36.

Finally, there is some good news. Tests have shown any wire under eighteen inches long is too short to act as an antenna to absorb E1 energy from an EMP due to the much higher frequencies involved. This means most handheld battery-powered devices including portable radios, cell phones, pagers, walkie-talkies, digital cameras, digital watches, iPad, and GPS devices are too small to be affected by an EMP. However, if these same devices are connected to a separate antenna, earphone cable, or wall charger, even if the charger is not plugged into an outlet, this will offer enough wire length to absorb the incoming E1 energy and damage the sensitive microelectronic devices inside.

Again, smaller battery-powered devices offer good resistance to EMP-induced high voltage surges, as long as they are not connected to any accessories or chargers at the time an EMP strikes. For

PROTECTING HOME APPLIANCES FROM EMP

maximum backup protection, you might consider having a spare backup device stored in a Faraday cage. While this may be impractical for a cell phone or pager used daily, it is good insurance for a battery-powered shortwave radio and a pair of walkie-talkies.

While you can utilize these suggestions to better protect your home's electrical wiring and electronic devices, it still could be of little use if the grid system supplying your home is down for six months or more. Remember, while an EMP can damage household appliances, the damage E3 and a solar storm can do to high-voltage utility grid transformers will be catastrophic, and could take up to a year to manufacture, ship, and install replacement transformers.

In other words, a more productive grid down goal is to find ways to live comfortably without all the major grid-powered household appliances and start acquiring those battery-powered devices that can replace those 120-volt AC electrical and electronic devices you cannot live without.

CHAPTER 34

Protecting Solar Systems from EMP

I described in chapter 28 the basics of how solar power can be used to charge a stand-alone battery and also recharge any battery-powered device. Other chapters address multiple ways batteries can be used to power emergency lighting, radios, and most small electronic devices.

Battery-powered devices may be the only electrical appliances and electronic devices still operational after a major grid down event, regardless of the cause. Without a working grid or generator, solar is the only remaining viable way for most people to keep their batteries fully charged. During my research I have "fried" multiple solar system components in real EMP test chambers and was very surprised by the results.

These tests proved without added safety devices, the E2- and E3-induced voltage surge feeding back through the grid or house wiring will destroy the sensitive microelectronic components in any interconnected solar inverter. In addition, E1-induced high voltages and currents can enter into a roof mounted solar array and all above roof solar wiring. This will damage the solar panels and destroy the microelectronic components in the solar charge controller wired between the solar array and battery bank.

SOLAR BATTERIES

An EMP will *not* damage deep-cycle open or sealed AGM or GEL batteries typically used to power golf carts, RVs, and boats.

PROTECTING SOLAR SYSTEMS FROM EMP

While these batteries are very common for residential off-grid solar and backup power systems, they are slowly being replaced by the much more efficient and higher capacity Lithium-Ion and Lithium Iron Phosphate batteries. While very expensive if purchased new, thousands of these tiny batteries assembled into "trays" are being replaced in older all-electric vehicles.

There are service centers refurbishing these used battery trays which includes replacing any defective individual battery cells. Even though they will be used, these vehicle tray battery assemblies still have many years of remaining life in off grid solar applications. While this new battery technology will eventually become the main battery technology for future off grid solar and backup power systems due to their lighter weight, longer life, faster recharge time, and much deeper discharge capability, there is one major flaw and that's the EMP risk.

All lithium-Ion vehicle tray batteries and the more temperature stable lithium iron phosphate batteries require a battery management system (BMS). These electronic devices equally distribute the electric flows and closely monitor battery temperatures throughout the multiple separate battery "strings" within each tray or self-contained battery box. Each vehicle tray battery typically contains over five hundred individual batteries approximately the size of a standard AA battery. These BMS units wired into each battery tray contain microelectronic components which would be damaged by an EMP event if the metal battery trays and interconnect wires do not provide adequate EMP shielding.

SOLAR ARRAYS

My EMP testing found that larger individual solar panels were damaged by an EMP, especially when their rear junction boxes were connected to long DC wires leading back to a solar charge controller

PROTECTING SOLAR SYSTEMS FROM EMP

or combiner box. I found an EMP can damage both the solar panels in a solar array, and the solar charge controller, so you need to protect all interconnecting DC wiring at both ends.

The easiest way to reduce the risk of E1-induced voltages and currents from damaging a solar array and solar charge controller is to install broadband ferrites, first introduced in chapter 33. These are installed at each end of the long DC wiring between the solar array to the solar charge controller.

However, unlike the installation of a separate ferrite on each AC power wire, for DC wiring both the positive (+) and negative (-) solar wires need to pass through the same ferrite in pairs. The wire from the solar array on the roof or in the yard to the charge controller inside the home makes a great "antenna" to absorb E1 energy. This is why attenuation devices need to be located at each end and for each solar panel.

Of course, there are always exceptions and complications, and in this case it's the latest revision to the National Electric Code (NEC). Although not yet adopted by all state and counties, the 2020 NEC requires all exterior solar wiring to be automatically disconnected at the above roof array upon either activation of a manual emergency disconnect or by some form of electronic signal from the inverter or charge controller. Solar manufacturers are not even sure how best to make this work, but most likely this will require additional communication or signal wires between the exterior solar array and interior solar power system.

This means additional cabling could also act as an antenna to send EMP-induced high voltages and currents down into the solar equipment so these wires will also need ferrite protection. In addition to installing ferrites at each end of the wire between the solar array and solar charge controller to block the higher frequency E1 component of an EMP, there are now DC rated surge suppressor

PROTECTING SOLAR SYSTEMS FROM EMP

devices to block the lower frequency E2- and E3-induced voltage and current surges in DC circuits.

Just like the 120/240-volt AC surge suppressors made to protect the incoming 120-volt AC grid power, these DC rated surge suppressors should be located at each end of the wiring supplying the solar charge controller. These basically are a small sealed box having positive (+) and negative (-) connectors on each side. These are typically only used with solar arrays having a large number of individual solar panels.

While there are several manufacturers now making surge suppressors designed specifically for DC power wiring, the Schaffner #FN220 series is a good all-around choice. Since the higher amp capacity units can be expensive, select a size that is about 25 percent larger than the maximum amp rating of your solar array. Midnight Solar also makes a very good high-quality lightning arrestors for both AC and DC circuits in a solar power system. These have less than a one microsecond response time and very high voltage blocking capacity. While originally intended as lightning protection for solar systems, these will still provide some protection against an EMP.[1]

SOLAR CHARGE CONTROLLERS

All solar arrays and even just a single solar panel require a solar charge controller wired between the solar power source and the battery being charged. Chapter 28 provided a brief description of each size and type, but all will be subject to EMP-induced high voltages and current surges that will travel down the wires from a rooftop solar array.

Larger solar charge controllers for whole-house solar systems can be very expensive and will benefit from these EMP protective measures. Smaller solar backup systems under 400 watts in size do not normally require a large solar charge controller, and the solar

PROTECTING SOLAR SYSTEMS FROM EMP

panels may actually cost less than the cost of the added EMP protection.

Fig. 34-1. MPPT solar charge controller before EMP exposure meter indicates its charging at 13.32-volts.

Fig. 34-2. MPPT solar charge controller after EMP exposure. Meter indicates it's now charging at 16.51-volts.

While properly sized ferrites can attenuate the induced E1 high voltages coming from the solar array into the solar charge controller, by far the best fail safe protection against EMP damaging a solar charge controller will be to just have a spare unit! Since most are fairly small and reasonably priced, I recommend keeping a spare solar charge controller in a Faraday protective device as discussed in chapter 36.

PROTECTING SOLAR SYSTEMS FROM EMP

SOLAR INVERTERS

My EMP testing destroyed several different brands and sizes of inverters typically being installed in today's off-grid solar systems. If you already have a whole-house solar power system or emergency battery backup system, there are several ways to harden your solar system against EMP. Even if you are not connected to the utility grid and are living a totally off grid lifestyle, you still may have conventual 120-volt AC wall outlets with plenty of wiring back to an inverter and perhaps an outside generator.

In chapter 33 I introduced the use of ferrites to protect the primary 240-volts AC feeds coming into your home's main circuit breaker panel from the electric meter and grid. These same ferrites should also be used on all 120/240-volt AC wires going into and out of an inverter and generator.

However, a battery inverter is only half of the solar home's electrical wiring exposed to an EMP. On the other side of the inverter will be heavy DC cables connecting a battery bank, a separate DC circuit breaker panel, a solar charge controller, plus long wires connecting a solar combiner box and multiple roof or ground mounted solar panels. Ferrites can attenuate EMP-induced high-frequency E1 voltage surges induced into these DC wires.

Fig. 34-3. One of several new inverters I tested in an EMP chamber. None survived.

PROTECTING SOLAR SYSTEMS FROM EMP

PACKAGED SOLAR SYSTEMS

There are a few inverter manufacturers starting to recognize the potential risks of both an EMP and a solar storm and are adding additional components to block high voltage and current surges entering their inverters and solar charge controllers.

It should be noted Sol-Ark manufacturers a complete solar power system package that includes an inverter specifically tested to withstand EMP-induced high voltages and currents. Their packaged systems include properly sized ferrites to install on the solar array, solar wiring, and solar charge controller. While intended for larger whole-house solar power applications, this is one of the most popular inverters specifically marketed as being EMP protected and the Sol-Ark products have undergone extensive testing in EMP test chambers.

If you are not considering a whole-house solar power system, I suggest keep it simple and use foldup solar panels to recharge all of your smaller electronic devices. Due to their small size when folded, these flexible solar panels will not be damaged by an EMP, as long as not connected to any long wires when an EMP strikes.

CHAPTER 35

Bugging Out after an EMP

Okay, now it's time. You fought the good fight and have stayed in your home as long as you could using a generator or solar backup system, but now you have to leave. It's usually better to stay in your own home during any emergency or grid down event as long as you have made the proper preparations and are not in immediate danger. But sometimes people are forced to evacuate during a disaster with only a few minutes of warning and may never be allowed to return. Watch the evening news, as this is happening somewhere in the United States every single day due to hurricanes, flooding, forest fires, mud slides, tornados, power outages, a tidal wave, or chemical spills.

This could also mean you will be on foot with only what you can carry in a backpack if an EMP has damaged your vehicle or out of gas due to closed service stations. While traveling in a well-stocked RV or truck camper would be ideal, this may not be possible after a real EMP attack or solar storm.

For example, you could be hundreds of miles away on vacation, away at school, or even in the middle of work-related travel. When a grid down event occurs, it may be far safer to immediately head for home or an alternate safe meeting place before traffic gridlock and gang violence begins. It's also possible all public transportation has already stopped, and driving is no longer possible.

BUGGING OUT AFTER AN EMP

Fig. 35-1. Small battery LED lights and radio are not affected by an EMP and should be in every bug-out bag.

A book bag or gym bag is a great way to carry the few survival items you will need, especially if you have to bug out from an office or apartment building in a large city. Nobody would expect survival supplies or anything of real value to be carried around in a book bag! However, a loaded-down military camouflaged backpack will not stay on your back long when you find yourself being overtaken by a panicked crowd rushing down the city street to escape some disaster.

There are endless magazine articles concerning what to pack in a bug-out bag, but I like to keep things simple. I would much rather have several identical backpacks holding the same basic supplies with one in each vehicle, than to have the perfect oversized backpack holding everything possible, but it's sitting at home fifty miles away when needed. Keep in mind, bug-out backpacks for emergency evacuation purposes are only to help you get from where you are now, to where you need to be. They are not full of everything you need for a two-week long camping trip, so keep the size and weight to a minimum!

Each bug-out bag I have includes some bottled water, a water purifier bottle, several high-energy snack bars, two dust masks, a vacuum packed rain slicker with hood, thermal space blanket, quality LED headlight with head strip and extra batteries, one medium size folding knife, parachute cord, roll of duct tape, fold-up street map, Bic lighter, long-handled stainless steel coffee cup, canned heat (easily fits inside coffee cup), fire-starter gel packs, referee whistle, travel size toilet paper, travel size soap, travel size toothpaste and toothbrush, foot powder, bandages in various sizes, a tourniquet, and small mirror.

As a side note, I am often asked to speak at self-preparedness events along with other speakers addressing other areas of being prepared. One of the speakers I met was a very well-known expert featured on several TV shows who starts campfires, typically by rubbing sticks together, striking a flint on a rock, and other primitive and very time-consuming fire-making techniques. I asked him while we were backstage with nobody around how he normally started a fire when he was alone. He paused, looked around, then said he always carries a Bic lighter!

Years ago, many people smoked and almost everyone carried a flip-top metal lighter or book of matches. Since most kitchen stoves were either gas or wood fired, every kitchen had a large box of wooden kitchen matches. However, today there are few smokers, and kitchen stoves are either instant start gas or all electric. Ever try to light the candles on a kid's birthday cake and nobody had a match? If you are ever stranded in a vehicle or having to walk for days, being able to start a fire to heat a cup of water to make a broth or coffee will save your life.

Space and weight will be at a premium, but you should still have room for a battery-powered handheld radio and earphone. You could find yourself on foot after your vehicle ran out of fuel, was damaged by an EMP, or you just lost control on ice and slid into a ditch with nobody around. You will still need your cell phone to function or you will soon be lost, in the dark, and will not know what is beyond your immediate location. A foldup solar charger with 10-watt output is perfect to keep your cell phone, GPS receiver, and LED flashlight charged, and these are small enough to carry if you are on foot.

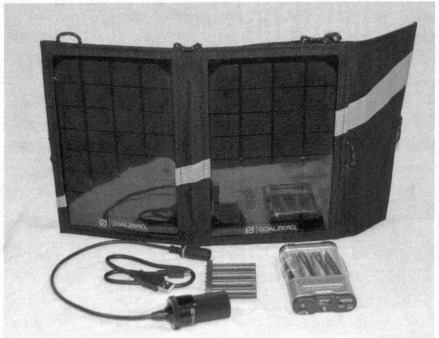

Fig. 35-2. Fold-up solar panel and charger for four AA batteries.

Beside each bug-out bag I keep a pair of really good hiking boots and two pairs of hiking socks. These are not tennis shoes and sweat socks. These were selected for having the right design for miles of walking, which may be the only form of transportation available when a grid down event first hits. I applied multiple applications of a quality boot waterproofing and wore each pair several days to break them in before storing in the trunk with each bug-out bag.

The last thing you need during an emergency is to hike twenty miles in a brand-new pair of shoes or boots! I am sure other texts will suggest a different combination of must-have emergency supplies, but this bag and hiking shoes are just to get you to a much safer place and are not intended to support a weeklong camping trip.

A true emergency evacuation may require walking many miles when it could be cold, raining, and dark. Don't assume an evacuation will involve a sunny spring day on flat ground as you walk along admiring the flowers and nice scenery! Many pre-assembled survival bags are filled with inexpensive gadgets just too look impressive, but they usually have little or no use in a real emergency. I suggest making up your own bug-out bags with this book as a starting point.

Make sure you switch the protective clothing stored with each bug-out bag in early spring and again in early fall to account for the seasonal changes. For example, pack a pair of insulated gloves and heavier hooded jacket just before winter, then switch to a lightweight jacket and hooded rain slicker in the spring. An umbrella is a waste of space and will quickly fail in a high wind so make sure all coats and jackets include a waterproof hood. An extra bottle of water may come in handy during summer months but could freeze in the trunk of a car during winter months. Energy bars and some of the other suggested supplies have expiration dates and should be replaced each year as needed.

CHAPTER 36

Basic Faraday Devices for EMP Protection

Even when you are inside a metal enclosed elevator, inside a metal framed building, surrounded by other concrete and steel highrise buildings, you can still pick up an AM radio broadcast on a handheld radio. Your cell phone will also ring, indicating an incoming call even if you are not able to complete the call due to partial signal attenuation by the structure. If these extremely weak radio transmissions can reach your electronic devices even here, there is no question a huge electromagnetic "shockwave" containing all frequencies and magnified several million times will reach every electronic device you own.

While not as good as the shielding provided by a solid metal enclosure, common building materials will still provide a level of EMP shielding across the higher frequencies associated with E1. Since E1 electromagnetic energy can enter and damage electronic devices without needing the power and data wires that carry E2 and E3 voltage surges, partially blocking this radiated E1 energy will at least reduce the potential damage to electronic equipment located inside any structure.

Electromagnetic shielding effectiveness, or attenuation, is measured in "decibels" which is a logarithmic scale. For example, a 40 dB reduction is 100 times more effective shielding then 20 dB, while 60 dB is 1,000 times more effective than 20 dB and an 80 dB reduction is 10,000 times more effective than a 20 dB reduction, and

BASIC FARADAY DEVICES FOR EMP PROTECTION

so on. For reference, shielding that provides over 50 dB of attenuation access all frequencies of an EMP will almost guarantee no damage to electronic devices behind this protection.

While relying on just the building structure alone to protect critical electronic systems from EMP is a fool's errand, testing does indicate electronic equipment in every commercial and institutional building will not suffer EMP damage equally. It's realistic to expect a heavy masonry or all metal building with minimum windows could provide enough E1 attenuation to allow only minor system damage that is repairable, verses catastrophic system destruction for no protection at all.

For example, while a standard 2 x 4 wood exterior residential wall with glass windows are both transparent to E1, heavier wall construction does offer a fair level of E1 signal attenuation.[1] This includes:

Wall Material	dB Attenuation*
8-inch hollow concrete block	12 dB
Brick over 8-inch hollow block	14 dB
16-inch hollow concrete block	17 dB
8-inch poured concrete	23 dB
12-inch poured reinforced concrete	35 dB

* At 900 mHz

Of course, this attenuation will vary depending on the frequency of the signal. Earlier chapters describing my actual EMP testing have shown handheld electronic devices including battery-powered cell phones, portable radios, and walkie-talkies are too small to couple with electromagnetic EMP energy even at very high energy levels. Testing has shown it takes a wire or antenna over eighteen inches

BASIC FARADAY DEVICES FOR EMP PROTECTION

long to couple with this energy and transfer the resulting high-voltage and currents this EMP energy generates into the microelectronic components inside.

Obviously, it's not practical to keep your cell phone inside an EMP protective enclosure, especially since there is limited risk to a device this small, as long as it was not plugged into an earphone or charger when an EMP event happens. However, just to be extra careful, it's not unreasonable to have a spare battery-powered shortwave radio, calculator, DVD player, walkie-talkies, a spare control board for your generator, a spare solar charge controller, and other useful electronic devices stored in some form of EMP shielding. Just remember to never leave batteries in any electronic device when stored for long periods as they all eventually leak and corrode the battery contacts inside the device, regardless of advertising claims!

We credit the design of any enclosure used to block EMP energy to Michael Faraday, a brilliant and self-educated scientist living in England during the mid-1800s. His experiments discovered the relationship between the flow of electricity through wires and the resulting magnetic fields, which became the basis of all electromagnetic theory in physics today. His research also proved electromagnetic energy travels along the surface of a conductor, not inside. He found any hollow enclosure made from an electrically conductive material will absorb or block any electromagnetic energy that strikes the exterior, keeping it from reaching any objects placed inside.

However, the exterior of the enclosure will still be electrically energized, so there needs to be some form of nonconductive insulation covering all interior surfaces to prevent contact with the stored items. New metal gallon paint cans with tight fitting metal-to-metal lids with added cardboard or plastic covering the inside surfaces will provide excellent EMP protection for small electronic

BASIC FARADAY DEVICES FOR EMP PROTECTION

devices, including a spare control board for your generator. These are inexpensive and available from any builder supply store. You must have a tight metal-to-metal fit between the lid and the container for any enclosure you use to block EMP energy.

FOIL WRAPPING

Larger equipment containing micro-electronic circuits can be protected by multiple separate layers of heavier restaurant-grade aluminum foil. This is usually easier to do if the equipment is left in its original rectangular shipping packaging and the foil applied like wrapping a birthday gift. Tests have shown wrapping the box with three separate layers of restaurant grade aluminum foil will block most medium to high frequency electromagnetic energy, but it takes five separate layers to block out all EMP frequencies at the highest levels of exposure.[2] Care must be taken to seal over any tears or punctures in the foil during the wrapping process, and wrap and tape each layer separately.

Testing for EMP shielding effectiveness has been done on fire safes, static bags, ammo cans, disabled microwave ovens, garbage cans, and foil covered boxes. A 50-dB level of shielding is considered excellent protection when dealing with EMP,[3] but EMP enclosures provide different levels of protection depending on the frequency of the EMP energy field.

TRASH CANS

A metal garbage or trashcan with metal foil taped around the lid joint and containing an electronic device also enclosed in a electrostatic bag, will provide in excess of 50-dB of EMP shielding across the entire radio frequency spectrum from under 100kHz up to over 1 GHz. The following table shows how other common EMP barriers compare to this maximum goal. While most protective

BASIC FARADAY DEVICES FOR EMP PROTECTION

methods did fairly well at lower frequencies, the protection dropped off significantly at the higher frequencies contained in a real EMP.[4]

Shielding Effectiveness Table[5]

Frequency	Fire Safe	Static Bag	Ammo Can	Microwave Oven	Foil Box Covered	Taped Lid Garbage Can	Taped Garbage Can with Static Bag
100 kHz	35	40	>50	>50	>50	>50	>50
500 kHz	39	37	>50	>50	>50	>50	>50
1 MHz	37	41	>50	>50	>50	>50	>50
5 MHz	32	22	>50	>50	>50	>50	>50
10 MHz	22	19	>50	>50	>50	>50	>50
50 MHz	23	22	41	48	>50	>50	>50
100 MHz	23	15	26	34	>50	>50	>50
250 MHz	16	17	21	24	>50	>50	>50
500 MHz	15	18	38	45	21	41	>50
1 GHz	6	18	9	28	19	33	>50

For under $25.00 you can build your own EMP-proof container. I keep a battery-powered shortwave radio, multi-battery charger, several GMRS walkie-talkies, think pad computer, small portable TV, DVD player, calculator, and GPS unit in a five-gallon plastic bucket, placed inside a six-gallon metal trash can having a tight-fitting metal lid. There is still room for extra rechargeable batteries, spare internet router, and several sizes of fold-up solar chargers. I fill the empty space between the walls of the inner bucket and the outer trash can with spray foam "crack" insulation. This homemade container keeps out rain, heat, cold, moisture, chewing rodents, and as a bonus is EMP proof!

BASIC FARADAY DEVICES FOR EMP PROTECTION

Fig. 36-1. Plastic bucket inside galvanized trash can makes a great Faraday Cage.

Fig. 36-2. Sealing trash can lid with metal duct tape is required as EMP energy will pass through any unsealed joints.

After the lid is closed tight, for added safety wrap a layer of two-inch wide metal-foil tape around the joint which is normally used to seal residential ductwork. Since many of the battery-powered devices recommended in earlier chapters may not be needed during normal day-to-day activities, storing these spare devices in EMP and moisture-proof metal containers is low cost insurance to keep them safe if disaster strikes, and easy to grab if you have to bug out later. I actually have several to store different classes of devices and label the lid with a black marker pen to identify what is inside.

BASIC FARADAY DEVICES FOR EMP PROTECTION

STATIC BAGS

Originally sold to prevent static electricity from damaging shipped microelectronic parts, some static bags are now being sold as EMP barriers and state they meet all government EMP shielding standards. While the materials used to make these static bags may have a similar level of signal attenuation, there were major differences in signal attenuation test results due to how the bags are sealed closed.

Those having a method of folding over and sealing the opening did far better overall at the higher frequencies associated with E1 than static bags with a simple press seal closer typically used for food storage bags. Those bags advertised to meet stricter military standard #MIL-PRF-81705D or MIL-PRF-131 preformed much better at all frequencies above 500 kHz than the zip top closure static bags.[6]

Claiming static bags exceed military specifications can be misleading as the most widely referenced government specification #MIL-STD-188-125 and its recent updates are primarily intended to specify the procedures to test an EMP hardened room or mobile military communication systems and was never intended to specify EMP testing to determine the survivability of non-military EMP protection devices.

The most interesting result for all of these different shielding techniques is they all provided an almost equal

Fig. 36-3. Various mylar type bags specifically made to block RFI and EMP electromagnetic energy.

BASIC FARADAY DEVICES FOR EMP PROTECTION

level of shielding protection in the lower frequencies typically produced during the E2 and E3 phase of an EMP event. However, only the trash can with added static bag inside provided blocking of all higher frequencies found in the E1 phase of an EMP event.

BLACK BOX

Some of the "black box" devices recently entering the market claim to protect all of the electrical appliances and electronic devices in your home by just attaching the device to the main electrical house panel. I have had sales staff for some of these magic boxes claim I just needed to plug one into any wall outlet and it will protect every appliance in the house! Most are basically lightning surge suppressors and sell for hundreds of dollars more if they have an EMP proof sticker on the cover. Most of these devices just contain one or more metal oxide varistors (MOV) plus a gas-discharge surge arrestor as their only components to block or limit the high voltage that will be induced into the power lines entering your home from a lightning storm.

Tests described in chapter 32 have shown a MOV and gas-discharge surge arrestor will block the E2- and E3-induced high voltages and currents, but cannot switch fast enough to block E1, and cannot block any of the E1-induced voltage surges entering into all downstream house wiring and appliances.

Remember, the initial high frequency E1 portion of an EMP passes directly through walls, roofs, and non-metal equipment enclosures. The E1 phase of an EMP event does *not* need any electrical wiring to do its damage, so installing some type of whole-house surge suppressor upstream at the main house electrical panel does not block E1. Consider, having a tabletop 120-volt AC radio tuned to a strong AM radio station. The radio still receives the station's electromagnetic radio waves sent out from the station's

BASIC FARADAY DEVICES FOR EMP PROTECTION

broadcast antenna traveling through the air, regardless if the radio is plugged into a surge suppressor in the power cord or not. The frequency of this radio station is just one of the millions of other frequencies contained in an EMP.

CONCLUSIONS

While you are considering which existing 120-volt AC electrical appliances and electronic equipment to cram into the very limited space for most of these EMP protective measures, the real question is how will you power these devices if the electric grid is down for six months? This sobering fact should drive home my belief in having smaller battery-powered versions of any electrical appliances and electronic devices that you just must have when there is no electric grid or backup generator still operating. This should drastically reduce what items you really need to store in some type of EMP protection enclosure.

CHAPTER 37

Faraday Rooms for EMP Protection

Most readers are interested in providing EMP protection for only a few small electronic devices which can be accomplished with the methods I discussed in chapter 36. For larger equipment or multiple cases of stored electronic hardware, some readers may want to convert an entire room of their home, basement, or garage into an EMP shielding room as discussed in this chapter.

Commercial EMP shielded rooms are enclosed on all sides, top, and floor with a multi-layer "sandwich" of two solid metal sheets that have strong magnetic shielding properties, separated by a non-metallic insulating core at least half-inch thick. These annealed metal sheets come in various thicknesses which are attached to the room walls using non-metallic fasteners. Access should be limited to a single door, which includes overlapping copper finger strips around all perimeter openings. Magnetic shielding is actually slightly different from radio frequency shielding, as magnetic shielding improves as the thickness of the metal shielding is increased. This is one reason why commercially manufactured magnetic shielding metal wall materials are available in multiple thicknesses.[1]

A more practical shielding room can be made using copper screen mesh, which will block most of the electromagnetic energy from an EMP. In this case, a single layer of copper screen will work, as long as the openings in the mesh are small, and careful installation efforts are taken to overlap and seal all joints and corners with metallic copper tape. Normally traps are installed at any wiring or

FARADAY ROOMS FOR EMP PROTECTION

piping penetrations entering a shielded room. Typically, these are metal boxes provide one or more ninety-degree turns in the wiring or piping. In addition, all power or control wires must pass through high-voltage surge suppressor devices located inside these metal traps.

Fig. 37-1. Restaurant grade aluminum foil to line walls and door will turn any unused closet into a large Faraday Cage.

FARADAY ROOMS FOR EMP PROTECTION

It's also possible to turn a small closet into a Faraday cage. This will first require disconnecting and covering any wall switch or wall outlet with restaurant-grade aluminum foil, and removing any surface light fixture, and cover all openings with this foil. There should be no air vents or window openings. All interior surfaces, ceiling, floor, and interior side of the door must be covered completely using restaurant-grade aluminum foil which is less likely to tear.

All foil seams should be overlapped at least two-inches and covered with aluminized metal duct tape. Care must be taken to make sure the perimeter openings around the door frame and door threshold are also protected. This can be accomplished with rolled up aluminum foil inserted around the door from the outside after it is closed and locked. A battery-powered LED lantern hung from the ceiling will provide lightning for closets and smaller EMP rooms since these do not require power wiring.

EMP shielding attenuation in excess of 50 dB for this homemade closet enclosure is possible for all frequencies up to 100 MHz, but this protection drops off fairly quickly above 250 MHz as you approach 1GHz.[2] After covering the floor with this aluminum foil, add a sheet of plywood to protect the foil on the floor. Any shelving should be free standing and made from nonmetal materials. Utilizing a static bag to protect each of the most sensitive electronic devices placed in this room will significantly improve the overall signal attenuation across all frequencies.

A metal shipping container can be converted into a very large EMP-proof room up to forty feet long by eight feet wide. While these have all-welded steel sides and roof which provide high EMP attenuation, the floor is solid hardwood planks and will need to first be covered by two layers of heavy restaurant-grade aluminum foil or a single layer of copper screen. All seams and the intersection with

FARADAY ROOMS FOR EMP PROTECTION

the base of the walls should have at least a two-inch overlap and metallic tape sealed.

Fig. 37-2. New EMP shielding layer being covered with plywood flooring being installed in metal shipping container.

Normally a layer of plywood or a false floor consisting of two-by-four supports under the plywood will protect the metal sheet or copper mesh first placed over the original hardwood floor. A two-inch-wide copper tape should be used to cover each overlapping screen joint and seams between each wall, ceiling, and floor. While the adhesive backing on the copper tape does slightly insulate the outer copper foil from the copper screen its adhered, this does not reduce the joints shielding effectiveness.

While the doors are all metal, the rubber door gaskets will need to be replaced or incorporate some form of flexible copper fingers. Any totally shielded room large enough for people to enter needs to include a ventilation system to provide breathable air. The small air vents located near one end will need to be sealed similar to sealing the floor. If these existing openings will be used to incorporate some type of ventilation system, the opening should include an EMP duct filter as shown.

FARADAY ROOMS FOR EMP PROTECTION

Tests have shown a metal honeycomb section of ductwork at the point ventilation air needs to pass through a shielded wall will block most E1 energy from passing through. The individual A and B opening dimensions as shown in the following drawing must be less than 3½-inches each axis, with a minimum 26-inch long duct length. The metal flange surrounding this duct section must be welded to the metal container wall. The total number of these individual openings can be whatever is needed to accommodate the air flow and pressure drop required by the ventilation fan. The ductwork attached to each end of this EMP duct filter can be made using any typical duct materials and duct assemblies.

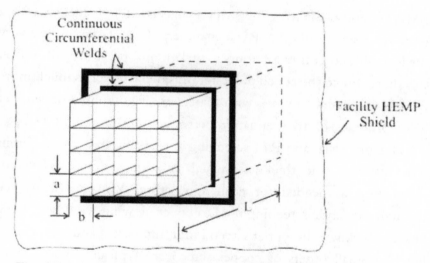

Fig. 37-3. *Metal honeycomb at wall penetration allows any size air ducts to enter and exit a large Faraday room.*

It is recommended that any computer or phone communication cables be converted first to fiber optic cables before penetrating any walls of an EMP shielded room. While the fiber optic converter on the outside of the EMP shielded room will most likely be damaged and need replacement in the event of a real EMP, the fiber optic

FARADAY ROOMS FOR EMP PROTECTION

cables will block all high voltages and currents in communication cables from passing into the electronic equipment located inside the shielded room. Any power wiring that must penetrate into an EMP shielded room should include both MOV and ferrite filtering devices at the point the wires enter.

A Faraday cage does not need to be grounded to maximize its shielding effectiveness, but grounding can be used to bleed away any built-up charge, especially for the very high-frequency saturation that can occur in room-size Faraday cage shielding.[3] Grounding is *not* recommended for the smaller EMP protection devices like a trash can or other protective methods discussed in chapter 36.

I have used flexible metallic fabric screening to make several EMP coverings for irregular shaped objects and to shield viewing window openings into EMP test chambers. These fabrics have a very small mesh and consist of copper, nickel, and polyester, all woven together. Due to the added polyester, this metal screen is much more flexible and easier to work with than more rigid copper screen, yet it still provides 45 dB attenuation between 30 mHz and 8 GHz.

Flexible EMP and RFI screening is available in many different styles, weaves, and thickness, which provide a different level of attenuation as needed for each application. You can easily see through the fabric screening used to cover viewing windows, while heavier thickness all-copper screen is not transparent and is much less flexible. Small orders of copper screen are typically folded up to reduce shipping costs and are very hard to flatten back out when received, so make sure your orders are large enough to guarantee they are wrapped around a cardboard tube before shipping. I have had to trash two separate copper screen orders that were folded over so many times I could not remove the heavy wrinkles in order to obtain a good seal around wall perimeters!

FARADAY ROOMS FOR EMP PROTECTION

Fig. 37-4. All copper mesh screen, door seals, and tape are available from multiple manufacturers.

For those of you interested in making these larger do-it-yourself EMP rooms or stand-alone EMP and RFI shielded chambers, there are multiple sources of parts and materials you can find on the internet. Some manufacturers specialize in providing metal wall sheeting that has much better magnetic shielding properties than aluminum foil or galvanized sheets. Other firms specialize in supplying many sizes and shapes of metallic door shields, while still others offer flexible copper and metallic fabrics to provide EMP shielding. All of these materials are available with different shielding attenuation levels which affects pricing. Be sure to verify the attenuation and frequency range specifications for the materials you order.

CHAPTER 38
Closing Comments

Fig. 38-1. Plan for the worst, hope for the best!

In April 2003, over sixty-five senior executives representing all electric utilities, federal and state government agencies, and National Laboratory representatives came together to plan out the future of the United States electric grid. The final result of these meetings was the publication of "Grid 2030," a document that looks into what needs to be done during the next hundred years to take the nation's electric grid into the twenty-first century. The executive summary of this report states, "America's electric system, the supreme engineering achievement of the 20th century, is aging, inefficient, congested, and incapable of meeting the future energy needs of the

CLOSING COMMENTS

Information Economy without operational changes and substantial capital investment over the next several decades."[1]

While this report talks glowingly about the need to embrace everything from smart meters, flywheel energy storage, superconductor transmission lines, distributed energy, grid interacting appliances, hydrogen fuel cells, mini-grids, micro-grids, smart grids, and solar power, nowhere in this entire document was there even a mention of the vulnerability of the grid from an EMP or solar storm!

There was no mention how a coordinated terrorist attack on carefully selected high-voltage substations could bring down the entire electric grid serving over half of this nation and take years to restore. There was no mention that 85 percent of all high-voltage transformers purchased each year are imported; and it takes two-years to custom design, build, test, ship, and install. So much for trusting utility grid operators to take us into the twenty-first century!

This shows again how the electric utilities and those government agencies assigned to monitor the electric utility industry have had their combined heads in the sand, thinking if they do not mention the words *EMP* or *solar storm* they will never happen. There is a common human psychological condition called "normalcy bias" which causes people to grossly underestimate the likelihood that a disaster will occur, and minimize how bad the aftermath will be, simply because they believe things will always stay the way they have always been. In addition, the more unthinkable the consequences of something could be, the more they pretend it will never happen, so they will not need to even think about it!

I believe the entire management of the electric utility industry, plus a large part of the United States government have a normalcy bias regarding what an EMP or solar storm could do to the electric grid. While I am sure there are military planners and senior level government officials that do understand the risk, it's evident their

CLOSING COMMENTS

efforts have gone into the continuity of government (COG) planning to make sure they will survive, while letting everyone else fend for themselves.

This initial planning was just to have designated underground shelters for the president, congressional leadership, the supreme court, military brass, and all major department heads. It didn't take long for these officials to realize they just couldn't work without their immediate staff, so shelter plans and emergency supplies had to expand. Then some congressmen finally realized they were not going to live underground for the next six months while their wife and kids were stuck above ground and probably sleeping on a cot in a FEMA camp once their pantry was bare. So again, the underground shelter designs had to be expanded. You will note all this planning is for themselves and their families, not for you or me. I only hope they don't realize next week that they just can't leave Fluffy the cat and Buster the dog behind!

If you are reading this book, I think it's clear you consider a major grid down event to be a real possibility and do want to do more to prepare. You will be surprised to learn you are not alone, as there are a few million of us out there that do get it. The hardest people to convince are family members and friends. For years the prepper community played the part of Paul Revere only to be ridiculed for saying, "the sky is falling, the sky is falling," especially when Y2K didn't cause the end of the world as we know it! Most preppers today realize a large part of our society will never believe this country is in for some real hard times and see no need to prepare.

I have overheard some say they will just take what they need if things get really bad. Many claim they plan to just hunt when they need food. Perhaps they do not know that during the great depression there was not a single deer, rabbit, or squirrel to be found until years later since everyone with a shotgun had the same idea! Serious preppers are keeping a very low profile and have started to

CLOSING COMMENTS

form loose-knit groups of like-minded individuals who can band together if things turn really ugly.

Most preppers will go out of their way to help others who do not know how to begin. However, you have to initiate the contact, as they have found out the hard way how the nail sticking up gets hammered down. If you are just getting into prepping or want to know about EMP, I recommend attending one of the Preparedness or Self-Reliance Expos offered every few months in many parts of the country. You will be overwhelmed with all the preparedness products and the free educational classes available.

It really makes no difference if the coming grid down event will be caused by an EMP weapon, solar storm, failing infrastructure, natural disaster, sabotage, or just incompetence. The result will be almost the same. Your state, and perhaps your entire half of the United States could be without the electric grid for up to a year, and long past your generator's fuel supply. Both Hurricane Katrina in New Orleans and Hurricane Sandy in Long Island proved government relief agencies are incapable of a rapid aid response and could never provide enough emergency food, water, and shelter when many millions and millions of people are affected.

There have already been multiple acts of sabotage to isolated utility substations and transmission lines which are rarely reported by the national news media. These may be rehearsals for a future coordinated attack at multiple substation locations. With rapidly changing world events we can expect these acts to continue, and most likely impact far larger sections of our utility infrastructure in the near future.

Electric utilities are constantly expanding the existing grid systems to add new homes and businesses. However, a large number of the original sub-stations and high-voltage transmission lines supplying these new loads from distant generating plants were built in the 1960s and are way past their design life. As referenced in this book, there have also been numerous congressional hearings and studies highlighting the vulnerability of the electric grid from an EMP attack and solar storm, yet little if anything has been done by the utility operators to harden their systems against this potential major destruction.

While almost 40 percent of all power generating plants in the United States are still coal-fired, pressure from environmental groups and the EPA is forcing the closure of a large percentage of those remaining coal plants and there are no plans for their replacement. As discussed in chapter 5, over 552 perfectly operational coal-fired power plants have been scrapped. The average age of all fifty-eight nuclear power plants operating in the United States is forty years, and all are near the end of their design life. No doubt the national grid will be having capacity issues for years which makes us even more vulnerable if hit by an EMP.

Chapter 5 provided an introduction to the damage an EMP or solar storm can do to the nation's electric grid system, especially related to large high voltage transformers that take a year or more to build, deliver, and install. Chapter 6 provided a long list of the many congressional studies that have been completed that identified the risks and provided recommendations to reduce those risks from an EMP or solar storm. I noted how the utility industry and their lobbyists continue to drag their feet for making any system upgrades or adding EMP grounding to every high-voltage transformer and substation, which would also protect these systems from a solar storm.

It's important for us all to have a better understanding of the EMP recommendations included in the National Defense Authorization Act just signed into law by President Trump on December 20, 2019.

This act directs multiple government agencies to address the risk of an EMP attack to each specific agency and gives them a specific date when they must report back with their assessment of risk and what they plan to do about it. This is long awaited good news for those of us who have been warning for years about the dangers of an EMP. However, there is a major obstacle waiting in the wings that may block any chance to reduce our nation's vulnerability to an EMP attack.

No doubt this debate is going to soon boil over into the public square, and it's important for us to understand how the electric utility industry wants to avoid doing anything. It's also necessary for us to hold the electric utilities and our government representatives accountable if they do not. Identify those elected representatives pretending this is a non-issue and vote them out of office.

The Department of State, Department of Energy, Homeland Security, and the Nuclear Regulatory Commission will be the lead agencies to implement the recommendations listed in this EMP legislation, yet none of these agencies were part of the original EMP Commission. Most of their staff are holdovers from previous administrations that ignored the EMP Commission's recommendations for over twelve years, and many do not have a clue regarding EMP or solar storms. In addition, the electric power industry, acting through their own Electric Power Research Institute (EPRI), has already published multiple reports that claim the congressional EMP Commission's grid hardening recommendations are not needed as the EMP risks have been overblown!

Some of the EPRI reports claim a nuclear-generated EMP would be similar to the effects of a wide area hurricane, and easy to recover.[2] These utility lobbyists plan on a major campaign to convince the heads of these various federal agencies tasked with this EMP review they only need to consider grid protection against a solar storm. They insist more substantial hardening all high-voltage transformers against EMP is not practical, not needed, and an EMP will never happen.

Making the system upgrades to all high voltage transformers as recommended by the EMP Commission will guarantee total protection against a solar storm or Coronel Mass Ejection (CME).

CLOSING COMMENTS

Unfortunately, the reverse is not true. The minimum protections reluctantly being considered by the utility industry to protect against a solar storm are *not* adequate to provide protection against the E2 and E3 components of an EMP, which will definitely damage these very large transformers.

While I am encouraged the president has recognized the major threat to this country from an EMP attack, I am not optimistic his concerns will be followed through by the electric utility industry or the government agencies mandated to address these risks, and everyone will just kick the can down the road, again.

In closing, it's important to remember all small electronic devices powered by a battery and not plugged into a nearby wall outlet are almost guaranteed to *not* be affected by an EMP or solar storm. My 2016 book, *Lights On*, lists hundreds of battery-powered appliances and electronic devices that can substitute for any standard 120-volt AC appliance you need after the grid is down. However, if the only battery-powered devices you own are a cell phone and television remote, this would be a good place to start or your road to survivability!

I have shown how you can utilize solar power and battery-powered devices to make life less stressful regardless of how long it takes for the nation's electric grid to function normally again. When combined with other available reference books that address emergency food and water storage, home security, first aid, and weapons handling, this combined knowledge will help protect you and your family, regardless of what caused the coming grid down event or how long it will last. My favorite books on these related topics plus additional references about EMP are listed in the reference book section found in the appendix.

Stay safe and keep the Lights On!

APPENDIX

Recommended Reading

A Nation Forsaken (2013) by Michael Maloof
Blackout Wars (2015) by Dr. Peter Vincent Pry
Complete Guide to Disaster Preparedness (2014) by Scott Hunt
EMP – Hardened Radio Communications (2016) by William T. Prepperdoc
EMP Attacks and Solar Storms (2012) by Arthur T. Bradley, Ph.D.
EMP Manhattan Project (2018) by Dr. Peter Vincent Pry
EMP Survival (2011) by Larry and Cheryl Poole
Lights On (2016) by Jeffrey R. Yago
Lights Out (2015) by Ted Koppel
One Second After (2011) by William R. Forstchen
Pulse Attack (2016) by Anthony Furey
Raven Rock (2018) by Garrett M. Graff
Survive the End of the World as We Know It (2009) by James Wesley Rawles
Surviving EMP (2017) by Rob Hanus
The Grid (2016) by Gretchen Bakke, Ph.D.
The New Solar Electric Home (2008) by Joel Davidson
The Power and the Light (2020) by Dr. Peter Vincent Pry
The Survival Medicine Handbook (2016) by Joseph Alton, M.D.

Self Help Magazines

Self-Reliance Magazine
Backwoods Home Magazine
Mother Earth News Magazine

Photo Credits

Figure	Source
1-1	iStock by Getty Images 2020
1-2	iStock by Getty Images 2020
1-3	iStock by Getty Images 2020
1-4	iStock by Getty Images 2020
1-5	iStock by Getty Images 2020
3-1	"Late-Time (E3) High-Altitude Electromagnetic Pulse and its Impact on the Electric Grid" by Metatech Corp. report for Oak Ridge National Laboratory date 2010
3-2	"Late-Time (E3) High-Altitude Electromagnetic Pulse and its Impact on the Electric Grid" by Metatech Corp. report for Oak Ridge National Laboratory date 2010
3-3	"Late-Time (E3) High-Altitude Electromagnetic Pulse and its Impact on the Electric Grid" by Metatech Corp. report for Oak Ridge National Laboratory date 2010
3-4	"Late-Time (E3) High-Altitude Electromagnetic Pulse and its Impact on the Electric Grid" by Metatech Corp. report for Oak Ridge National Laboratory date 2010
3-5	"Late-Time (E3) High-Altitude Electromagnetic Pulse and its Impact on the Electric Grid" by Metatech Corp. report for Oak Ridge National Laboratory, dated 2010
5-2	iStock by Getty Images 2020
5-3	iStock by Getty Images 2020
9-1	iStock by Getty Images 2020
9-2	"Electromagnetic Pulse (EMP) Protection and Restoration Guidelines" by U.S. Department of Homeland Security, dated 2016
18-3	Tom Brennan, Sol-Ark Corp.
18-4	Tom Brennan, Sol-Ark Corp.
32-4	Dennis Bodson titled "Electromagnetic Pulse and the Radio Amateur" article, 1986 QST Magazine
33-3	Arthur T. Bradley,Ph.D. "EMP Attacks and Solar Storms" 2012
34-1	Tom Brennan, Sol-Ark Corp.
34-2	Tom Brennan, Sol-Ark Corp.
34-3	Tom Brennan, Sol-Ark Corp.
37-1	Arthur T. Bradley,Ph.D. "EMP Attacks and Solar Storms" 2012
37-3	"High-Altitude Electromagnetic Pulse Protection for Ground-Based Facilities", Department of Defense Standard MIL-STD-188-125-2, dated 1999
38-1	iStock by Getty Images 2020

Notes

Chapter 1: Understanding Electronic Disturbances

[1] Arthur T. Bradley, Ph.D., *Disaster Preparedness for EMP Attacks and Solar Storms*, pages 30-34.

[2] Arthur T. Bradley, Ph.D., *Disaster Preparedness for EMP Attacks and Solar Storms*, Table 4-9.

Chapter 2: What Is an EMP?

[1] Center for the Study of the Presidency and Congress, *Securing the United States Grid*, (July 2014), page 37.

[2] Dr. Peter Vincent Pry, *Blackout Wars*, (2015), page 35.

Chapter 3: EMP Basics

[1] United States Army Developmental Test Command, *Test Operations Procedure 1-2-622 Vertical Electromagnetic Pulse*, (September 2009).

[2] Oak Ridge Natural Laboratory, *The Lase-Time (E3) High Altitude Electromagnetic Pulse (HEMP) and Its Impact on the U. S. Power Grid*, (January 2010), pages 2-8 to 2-9.

[3] Oak Ridge National Laboratory, *The Late-Time (E3) High Altitude Electromagnetic Pulse (HEMP) and Its Impact on the United States Power Grid*, (January 2010), pages 2-4.

[4] United States Department of Homeland Security, *Electromagnetic Pulse (EMP) Protection and Restoration Guidelines for Equipment and Facilities (unclassified)*, (February 5, 2019), page D-11.

[5] United States Department of Homeland Security, *Electromagnetic Pulse (EMP) Protection and Restoration Guidelines for Equipment and Facilities (unclassified)*, (February 5, 2019), page D-13.

Chapter 4: Don't Forget the Sun

[1] Anthony Furray, *Pulse Attack,* Magna Carta Publications, (dated 2016), pages 16-18.

[2] Center for the Study of the Presidency and Congress, *Securing the United States Electric Grid*, (July 2014), page 70.

[3] F. Michael Maloof, *A Nation Forsaken*, World New Daily Books, (2013), pages 51-52.

Chapter 5: EMP Impact on the Power Grid

[1] Congressional EMP Committee, *Report of the Commission to Assess the Threat to the United States from Electromagnetic Pulse (EMP) Attack*, (April 2008), page 25.

[2] Staff of Congressman Edward Markey, *Electric Grid Vulnerability*, (May 2013), pages 7-8.

[3] United States Energy Information Administration (EIA) Washington, DC, (December 2019).

[4] United States Department of Energy, *Large Power Transformers and the United States Electric Grid*, (June 2012), pages 4-5.

[5] United States Department of Energy, *Large Power Transformers and the United States Electric Grid*, (June 2012), page 20.

[6] EMP Commission, *Report of the Commission and Assess the Threat to the United States from Electromagnetic Pulse (EMP) Attack*, (April 2008), page 33.

[7] North America's Reliability Corporation (NERC), *NERC 1989 Quebec Disturbance Report*, pages 51-52.

[8] Dr. Peter Vincent Pry, *EMP Manhattan Project*, (2018), pages 190-191.

[9] United States Department of Energy, *Large Power Transformers and The United States Electric Codes*, (2012), page 11.

[10] Federal Energy Regulatory Commission (FERC), *Internal Memo*, (March 2016).

Chapter 6: EMP Inaction by Government and Grid Operators

[1] Dr. Peter Vincent Pry, *Blackout Wars*, (September 2015), page 13.

[2] United States Congressional Commission, *Report of the Commission to Assess the Threat to the United States from Electromagnetic Pulse (EMP) – Volume One*, (2004).

[3] Congressional Commission, *Report of the Commission to Assess the Threat to the United States from an Electromagnetic Pulse (EMP) Attack*, (April 2008), Preface vii.

[4] Task Force on National and Homeland Security, *A Call to Action for America*, (October 2017), page 7.

[5] United States Department of Homeland Security, *The DHS Strategic Plan – Fiscal Years 2020-2024*, (December 2019).

[6] United States House of Regulations Report, *Electric Grid Vulnerability*, (May 21, 2013), page 4.

[7] Center for the Study of the Presidency and Congress, *Securing the United States Electric Grid*, (July 2014), page 18-19.

[8] Dr. Peter Vincent Pry, *Blackout Wars*, (2015), page 161.

[9] Dr. Peter Vincent Pry, *EMP Manhattan Project*, (2018,) pages 154-158.

[10] Dr. William R. Graham, *Chairman's Report to the Commission to Assess the Threat to the United States from Electromagnetic Pulse (EMP) Attack*, (July 2012), page 15.

Chapter 7: EMP Impact on Our Military

[1] Allison Maloney, *Broken Arrows*, UK the Sun article, (September 21, 2017).

[2] Major Colin Miller, *Electromagnetic Pulse Threats*, (November 2005), page 390.

[3] Dr. Ernest Allan Rockwell, *Electromagnetic Defense Task Force*, (2018), page 10.

[4] Dr. Ernest Allan Rockwell, *Electromagnetic Defense Task Force*, (November 2018), page 9.

[5] Taskforce on National and Homeland Security, *A Call to Action for America*, (October 2017), page 3.

[6] Dr. Gretchen Bakke, *The Grid* (2011), page 117.

[7] Dr. William R. Graham, *Chairman's Report to the Commission to Assess the Treat to the United States from Electromagnetic Pulse (EMP) Attack*, (July 2017), page 24.

[8] Dr. Peter Vincent Pry, *Blackout Wars*, (September 2015), page 61.

[9] Dr. William R. Graham, *Chairman's Report to the Commission to Assess the Threat to the United States from Electromagnetic Pulse (EMP) Attack*, (July 2017), page 21.

[10] Comrade General Chi Haotian, Vice Chairman of China's Military Commission, (December 2005), speech to top generals.

[11] F. Michael Maloof, *A Nation Forsaken*, (2013), page 39.

[12] Dr. Ernest Allan Rockwell, *Electromagnetic Defense Task Force Report*, (2018), Figure 1.5, page 17.

[13] Congressional and National Security Committee Hearings, *Threat Posed by Electromagnetic Pulse (EMP) to U. S. Military Systems and Civil Infrastructure*, (July 16, 1997), page 9-10.

[14] Task Force on National and Homeland Security, *A Call to Action for America*, (October 2017), page 7.

Chapter 8: EMP Effects on Communications

[1] National Coordinating Center for Communications (NCC), *Electromagnetic Pulse (EMP) Protection and Restoration Guidelines for Equipment and Facilities*, (December 2016), page D-4.

[2] Dr. William R. Graham, *Report of the Commission to Assess the Threat to the United States from Electromagnetic Pulse (EMP) Attack*, (April 2008), page 147.

[3] The President's National Infrastructure Advisory Council, *Surviving A Catastrophic Power Outage*, (December 2018), page 21.

[4] J. D. Heyes, *California to Ban Use of All Ham Radio Repeaters*, (October 14, 2019), Newstarget.com.

Chapter 9: EMP Impact on SCADA

[1] Center for the Study of the Presidency and Congress, *Securing the United States Electric Grid*, (July 2014), page 56-57.

[2] *Report of the Commission to Assess the Threat to the United States from Electromagnetic Pulse (EMP) Attack*, (April 2008), page 2.

[3] *Report of the Commission to Assess the Threat to the United States from Electromagnetic Pulse (EMP) Attack*, (April 2008), page 6.

[4] *Report of the Commission to Assess the Threat to the United States from Electromagnetic Pulse (EMP) Attack*, (April 2008), page 8.

Chapter 10: EMP Impact on the Banking Industry

[1] Congressional EMP Committee, *Report of the Commission to Access Threat to the United States from Electromagnetic Pulse (EMP) Attack*, (April 2008), pages 83-87.

[2] Congressional EMP Commission, *Report of the Commission to Assess the Threat to the United States from Electromagnetic Pulse (EMP) Attack*, (April 2008), pages 83-84.

Chapter 11: EMP Impact on Vehicles

[1] *Report of the Commission to Assess the Threat to the United States from Electromagnetic Pulse (EMP) Attack*, (April 2008), pages 115-116.

Chapter 12: EMP Effects on Transportation Control Systems

[1] Congressional EMP Commission, *Report of the Commission to Assess the Threat to the United States from Electromagnetic Pulse (EMP) Attack*, (April 2008), page 114.

[2] Federal Aviation Administration, *Air Traffic by the Numbers*, (2018).

[3] Major David Stuckenberg – Lemay Center for Doctrine, *Electromagnetic Defense Task Force 2018 Report*, (2018).

[4] Congressional EMP Commission, *Report of the Commission to Assess the Threat to the United States from Electromagnetic Pulse (EMP) Attack*, (April 2008), pages 111-112.

[5] United States Energy Information Administration, *Today in Energy* (2018).

Chapter 13: EMP Impact on Oil and Gas Distribution

[1] *Report of the Commission to Assess the Threat to the United States from Electromagnetic Pulse (EMP) Attack*, (April 2008), page 95-96.

[2] Report of the Commission to Assess the Threat to the United States from Electromagnetic Pulse (EMP) Attack, (April 2008), page 98.

Chapter 14: See Who Is at the Door
[1] Jeffrey R. Yago, P.E., Lights *On*, dated 2016, page 23.

Chapter 15: So, What Can We Do?
[1] Jeffrey R. Yago, P.E., *Lights On*, (2016), pages 15-16.
[2] Rick Martin, *Which is FEMA?* (September 2000), pages 1-2.
[3] Garrett M. Graff, *Raven Rock*, (January 2017), page 396.
[4] Garrett M. Graff, *Raven Rock*, (2017), page 128.
[5] Garrett M. Graff, *Raven Rock*, (2017), page 406
[6] Garrett M. Graff, *Raven Rock*, (2017), pages 407-408.
[7] Garrett M. Graff, *Raven Rock*, (2017), page 234.
[8] Harry V. Martin, *FEMA-The Secret Government*, (1995), page 4.

Chapter 17: Will Your Generator Survive an EMP?
[1] Anthony Furey, *Pulse Attack*, (2016), pages 72-73.

Chapter 18: Lighting after an EMP
[1] Jeffrey R. Yago, P.E., *Mother Earth News Magazine*, (March 2019), pages 14-15.

Chapter 19: Powering Communications after an EMP
[1] Jeffrey R. Yago, P.E., *Portable PV*, HomePower Magazine, (July 2015), page 28-37.
[2] Jeffrey R. Yago, P.E., *Amateur Ham Radio*, Backwoods Home Magazine, (October 2019).

Chapter 22: Powering Medical Equipment after an EMP
[1] Dr. Joseph Alton, *The Survival Medicine Handbook*, (2016).

Chapter 23: Powering Security Systems after an EMP
[1] Jeffrey R. Yago P.E., *Ice from the Sun*, Self-Reliance Magazine (Fall 2019).

Chapter 28: Charging Batteries with Solar Power
[1] Jeffrey Yago, P.E., *Lights On*, (2016), Chapter 21, "Building Your Own Solar Power Supply."

Chapter 29: Which Rechargeable Batteries?

[1] Jeffrey R. Yago, P.E., *Lights On*, (2016), page 42.

Chapter 32: Protecting Antennas and Shortwave Radios from EMP

[1] William T. Prepperdoc, *EMP-Hardened Radio Communications*, (2016), page 23.

[2] Dennis Bodson, Office of Technology and Standards, *Electromagnetic Pulse and the Radio Amateur*, QST Magazine, (November 1986), page 30-34.

[3] National Cybersecurity and Communications Integration Center, *Electromagnetic Pulse (EMP) Protection and Resilience Guidelines for Critical Infrastructure and Equipment*, (February 5, 2019), page 29.

[4] Dennis Bodson, Office of Technology and Standards, *Electromagnetic Pulse and the Radio Amateur*, QST Magazine, (November 1986), Figures 13, page 31.

Chapter 33: Protecting Home Appliances from EMP

[1] Dennis Bodson, Office of Technology and Standards, *Electromagnetic Pulse and the Radio Amateur*, QST Magazine, (November 1986), page 31.

[2] Arthur T. Bradley, Ph.D., *EMP Attacks and Solar Storms*, web page, www.disasterprepper.com.

Chapter 34: Protecting Solar Systems from EMP

[1] Midnight Solar Inc., *Solar Surge Device Installation Manual*.

Chapter 36: Basic Faraday Devices for EMP Protection

[1] National Cybersecurity and Communications Integration Center, *Electromagnetic Pulse (EMP) Protection and Resilience Guidelines for Critical Infrastructure and Equipment*, (February 5, 2019), page 72.

[2] Rob Hanus, *Surviving EMP*, (2017), page 48.

[3] Arthur T. Bradley, Ph.D., *Disaster Preparedness for EMP Attacks and Solar Storms*, (2012), page 21.

[4] Arthur T. Bradley, Ph.D., *Disaster Preparedness for EMP Attacks and Solar Storms*, (2012), Table 4-9.

[5] Arthur T. Bradley, Ph.D., *Disaster preparedness for EMP Attacks and Solar Storms*, (2012) Table 4-9, page 75.

[6] Dr. Arthur T. Bradley, Ph.D., *Disaster Preparedness for EMP Attacks and Solar Storms*, (2012), pages 77-78.

Chapter 37: Faraday Rooms for EMP Protection

[1] Magnetic Shield Corporation, *Magnetic Shield Rooms and Modular Enclosures*, Bensenville, Illinois

[2] Arthur T. Bradley, Ph.D., *EMP Attacks and Solar Storms*, (2012), pages 78-81.

[3] Arthur T. Bradley, Ph.D., *Disaster Preparedness for EMP Attacks and Solar Storms*, (2012), page 21.

Chapter 38: Closing Comments

[1] United States Department of Energy, Executive Summary, *Grid 2030 – A National Vision for Electricity's Second 100 Years*, (July 2003).

[2] Dr. Peter Vincent Pry, *Swamp Continues to Water Down Realities of EMP Threat*, (January 2, 2020), onenewsnow.com

INDEX

AC transformers, 23
Allegany Ballistic Laboratory, 138
Automatic Voice Network, 138
Azores Islands, 75
batteries
   AA, 112, 158, 161, 163, 188, 189, 192,
      194, 195, 226, 227, 229, 230, 231, 233,
      263, 272
   AAA, 147, 158, 163, 230
   C, 159, 163, 188, 195, 226, 227, 230, 231
   D, 163, 230
   lithium iron phosphate, 213, 233, 234,
      263
   NiCd, 229
   Ni-MH, 229, 231
battery management system, 112, 113, 233,
   234, 263
battery packs, 112, 113
BMS, 112, 113, 233, 234, 263
California Department of Forestry, 93
Carrington, 48, 49, 50
Carrington Event, 49, 50
Center for Security Policy, 88
charger
   solar, 175
chargers, 183, 197
   batter, 168
   battery, 198, 228, 230, 231, 234, 237, 238
   car, 164
   in-line, 237

solar, 165, 224, 225, 278
Cheyenne Mountain, 135, 137
CME, 45, 46, 47, 49, 50, 51, 86, 295
Coast Guard, 85
COCGCON Readiness Rating, 143
COG, 135, 137, 139, 141, 142, 292
COGCON, 142
Commission to Assess the Threat to the United States from Electromagnetic Pulse, ix, 64, 66, 67
Compton Effect, 38
Compton Recoil Electron, 38
Compton Scattering, 38
Compton, Dr. Arthur H., 39
Continuity of Government, 135, 137, 142
Continuity of Government Readiness Conditions, 142
Coordinating National Resilience to Electromagnetic Pulse, 65
Coronal Mass Ejection, 45
Coronavirus, 12, 142
cosmic rays, 18
COVID-19. *See* Coronavirus
Department of Defense, 63, 69
Department of Homeland Security, 64, 65, 68
Direct Digital Control, 97
E1, 30, 89, 196, 213, 245, 247, 248, 249, 250, 251, 256, 259, 260, 264, 266, 267, 275, 280, 287
E1 energy, 37, 44, 196, 208, 210, 213, 247, 250, 251, 255, 259, 260, 264, 274, 287
E1 phase, 281
E1 pulse, 89, 152, 245
E2 phase, 37, 42, 57, 92
E3, 42
E3 phase, 40, 51, 57, 71, 92, 281
Edison, Thomas, 47
Electric Power Research Institute, 73, 295
electrical grid, 29, 50, 63, 125
Electromagnetic Defense Task Force, 65, 83
electromagnetic fields, 27, 28
electromagnetic radiation, x, 18, 19, 22, 84, 102
electromagnetic wave, 24, 86, 90
Electromagnetic waves, 24

EMP simulator, 90, 102
EMP testing, 13, 100, 110, 113, 125, 147, 151, 208, 218, 243, 244, 255, 263, 267, 275, 280
Empress I, 86
Empress II, 86
Environmental Protection Agency, 71
Ethernet cables, 31, 89, 102
Family Radio Service, 165
Faraday cage, 245, 261, 285, 288
Faraday, Michael, 276
Federal Aviation Administration, 71
Federal Energy Regulatory Commission, 53, 67, 70, 71
Federal Nuclear Regulatory Commission, 71
Federal Reserve Board, 105
Federal Reserve System, 104
FEDNEC, 104
FEDWIRE, 104
FEMA, 16, 131, 132, 136, 137, 138, 139, 140, 141, 142, 205, 292
FERC, 53, 67, 70, 71
First Energy, 125
Food and Drug Administration, 71
Forstchen, William R., 126
Fukushima, 77, 79
G1, 50
G5, 50
Gaffney, Frank, 88
gamma rays, 18, 19, 25, 26, 30, 37, 38, 39, 40, 45
General amateur radio license, 170
General Mobile Radio Service, 166
generator
   whole-house, 130, 151, 152, 153, 200, 214
Global Positioning System, 117
GPS, 91, 111, 117, 149, 150, 171, 176, 177, 239, 260, 272, 278
Greenbrier Resort, 136
Grid Reliability and Infrastructure Defense Act, 67
Gulf of Mexico, 43, 69, 81, 83
ham radio bands, 170
ham radio operators, 93, 95, 247, 250
ham radio transmitters, 247, 249

Haotian, General Chi, 82
HEMP, 32, 44, 63, 69, 74
High Altitude EMP, 74
Homeland Security Administration, 68
integrated circuits, 32
ionosphere, 18, 19, 22, 24, 26, 27, 40, 41, 42, 43, 44, 76, 89, 171
kilohertz, 25
Kirkland Air Force Base, 85
lightning arrestors, 35, 247, 265
magnetic disks, 106
magnetic field, 20, 22, 23, 24, 26, 27, 28, 31, 40, 41, 43
mAh charge capacity, 231
Marconi, 47
Maunder Minimum, 46
megahertz, 25
Megavolt Amp, 56
metal oxide varistors, 152, 251, 255, 259, 281
microchips, 32, 90, 100, 112, 160
microelectronic devices, 31, 112, 260
microgrids, 125
microsecond, 34, 265
milliamp-hour (mAh) rating, 230
millisecond, 34
Morse Code, 169
Morse, Samuel, 47
Mount Pony, 138
Mount Weather, 136, 137
nanoseconds, 32, 34, 249
National Defense Authorization Act, 66, 294
National Guard, 131
National Oceanic and Atmospheric Administration, 50
National Security Telecommunications Advisory Committee, 105
National Weather Service, 190
NERC, 53, 70, 71, 72, 73, 102
Nine Mile Point, 54
NORAD Air Defense Command Center, 135
North American Electric Reliability Corporation, 53, 70
North Pole, 19, 20

nuclear bomb, x, 23, 29, 30, 32, 34, 37, 39,
    40, 42, 44, 68, 69, 72, 74, 75, 81, 86, 89,
    134, 136, 153
oil refineries, 122
One Second After, 126
pneumatic tubes, 96
Point Patience Navel Center, 86
Pry, Dr. Peter, 64
pump stations, 15
radio, 29, 34, 88, 90, 91, 111, 144, 145,
    166, 168, 169, 191, 193, 250, 258
  2-meter, 170, 251
  AM, 92, 162, 190, 192
  AM station, 26, 190
  amateur community, 94, 241
  antennas, 117
  battery-powered, 162, 223, 230
  CB, 166, 167, 168, 170, 251
  clubs, 94
  communications, 83, 117, 118
  frequency interference. *See* RFI
  FRS, 165
  frequency shielding. *See* shielding
  frequency spectrum, 277
  ham, 20, 89, 92, 93, 152, 169, 170, 171,
      250, 252, 253, 260
  handheld, 165, 271, 274
  interface, 86
  navigational beacons, 116
  NWS, 191, 192
  portable, 153, 163, 226, 238, 239, 260,
      275
  portable weather, 192
  shortwave, 25, 261, 276, 278
  stations, 76
  transmission, 47
  transmitters, 29, 47
  volunteer networks, 94
  waves, 20, 26, 40, 76, 171, 281
  weather alert, 92
radioactive fallout, 76, 87
radioactive isotopes, 79
radioactive water, 79
Raven Rock Facility, 136
Red Cross, 186
relays, 59, 96, 97, 105

RFI, 86, 110, 117, 243, 259, 288, 289
San Diego County Water Authority, 101
SCADA, 96, 98, 99, 100, 101, 102, 103, 115, 119, 120, 121, 122
Scorpion, 75
SCUD missiles, 75, 76, 83, 84
Secure High-Voltage Infrastructure for Electricity from Lethal Damage, 68
shielding, 27, 28, 103, 110, 119, 138, 183, 243, 246, 263, 274, 276, 277, 280, 283, 285, 286, 288, 289
solar charger, 159, 164, 174, 182, 185, 212, 225, 231, 232, 235, 272
solar panels, 174, 175, 199, 218, 223, 240, 262, 263, 265, 266, 267, 268
solar power system
    whole-house, 222
solar systems
    whole-house, 265, 267, 268
South Pole, 20
Starfish Prime, 29
steam power, 96
Supervisory Control and Data Acquisition, 98
surge arrestor, 224, 249, 250, 258, 259, 281
surge suppressor
    whole-house, 254, 256, 259, 281
surge suppressors
    whole-house, 255
SWIFT, 105
Technician class radio license, 170
TEPCO, 79
Tesla, 112
Tesla, Nikola, 47
Trestle, 85
United States Energy Infrastructure Administration, 119
walkie-talkie, 87, 153, 165, 168, 251
War Relocation Authority, 141
waveform, 251, 258
X-rays, 18, 19, 25, 26, 37, 40, 45, 46, 74

CPSIA information can be obtained
at www.ICGtesting.com
Printed in the USA
LVHW011946020920
664818LV00007B/1199